THE GREAT PRETENDER

THE GREAT PRETENDER

THE UNDERCOVER MISSION THAT CHANGED OUR UNDERSTANDING OF MADNESS

SUSANNAH CAHALAN

THORNDIKE PRESS
A part of Gale, a Cengage Company

GALE
A Cengage Company

Copyright © 2019 by Susannah Cahalan, LLC.
Permissions are listed on page 675–677.
Thorndike Press, a part of Gale, a Cengage Company.

ALL RIGHTS RESERVED
The publisher is not responsible for websites (or their content) that are not owned by the publisher.
Thorndike Press® Large Print Nonfiction.
The text of this Large Print edition is unabridged.
Other aspects of the book may vary from the original edition.
Set in 16 pt. Plantin.

**LIBRARY OF CONGRESS CIP DATA ON FILE.
CATALOGUING IN PUBLICATION FOR THIS BOOK
IS AVAILABLE FROM THE LIBRARY OF CONGRESS**

ISBN-13: 978-1-4328-7891-7 (hardcover alk. paper)

Published in 2020 by arrangement with Grand Central Publishing, a division of Hachette Book Group, Inc.

Printed in Mexico
Print Number: 01   Print Year: 2020

For the ones who need to believe

"You'd have to be crazy to get yourself committed to a mental hospital."

— *The Shock Corridor,* 1963

CONTENTS

PREFACE

The story that follows is true. It is also not true.

This is patient #5213's first hospitalization. His name is David Lurie. He is a thirty-nine-year-old advertising copywriter, married with two children, and he hears voices.

The psychiatrist opens the intake interview with some orienting questions: *What is your name? Where are you? What is the date? Who is the president?*

He answers all four questions correctly: *David Lurie, Haverford State Hospital, February 6, 1969, Richard Nixon.*

Then the psychiatrist asks about the voices.

The patient tells him that they say, "It's empty. Nothing inside. It's hollow. It makes an empty noise."

"Do you recognize the voices?" the psychiatrist asks.

13

"No."

"Are they male or female voices?"

"They are always male."

"And do you hear them now?"

"No."

"Do you think they are real?"

"No, I'm sure they're not. But I can't stop them."

The discussion moves on to life beyond the voices. The doctor and patient speak about Lurie's latent feelings of paranoia, of dissatisfaction, of feeling somehow less than his peers. They discuss his childhood as a son of two devout Orthodox Jews and his once intense relationship with his mother that had cooled over time; they speak about his marital issues and his struggle to temper rages that are sometimes directed at his children. The interview continues on in this manner for thirty minutes, at which time the psychiatrist has gathered nearly two pages of notes.

The psychiatrist admits him with the diagnosis of schizophrenia, schizoaffective type.

But there's a problem. David Lurie doesn't hear voices. He's not an advertising copywriter, and his last name isn't Lurie. In fact, David Lurie doesn't exist.

■ ■ ■ ■

The woman's name doesn't matter. Just picture anyone you know and love. She's in her mid-twenties when her world begins to crumble. She can't concentrate at work, stops sleeping, grows uneasy in crowds, and then retreats to her apartment, where she sees and hears things that aren't there — disembodied voices that make her paranoid, frightened, and angry. She paces around her apartment until she feels as if she might burst open. So she leaves her house and wanders around the crowded city streets trying to avoid the burning stares of the passersby.

Her family's worry grows. They take her in but she runs away from them, convinced they are part of some elaborate conspiracy to destroy her. They take her to a hospital, where she grows increasingly disconnected from reality. She is restrained and sedated by the weary staff. She begins to have "fits" — her arms flailing and her body shaking, leaving the doctors dumbstruck, without answers. They increase her doses of anti-psychotic medications. Medical test after medical test reveals nothing. She grows more psychotic and violent. Days turn into weeks. Then she deflates like a pricked balloon, suddenly flattened. She loses her ability to read, to write, and eventually she stops talking, spending hours blankly staring at a television screen. Sometimes she grows agitated and her legs dance in crooked spasms. The hospital decides that it can no longer handle her, marking her medical records with the words TRANSFER TO PSYCH.

The doctor writes in her chart. Diagnosis: schizophrenia.

The woman, unlike David Lurie, does exist. I've seen her in the eyes of an eight-year-old boy, an eighty-six-year-old woman, and a teenager. She also exists inside of me, in the darkest corners of my psyche, as a mirror image of what so easily could have happened to me at age twenty-four, had I not been spared the final move to the psychiatric ward by the ingenuity and lucky guess of a thoughtful, creative doctor who pinpointed a physical symptom — inflammation in my brain — and rescued me from misdiagnosis. Were it not for that twist of fortune, I would likely be lost inside our broken mental health system or, worse, a casualty of it — all on account of a treatable autoimmune disease masquerading as schizophrenia.

16

The imaginary "David Lurie," I would learn, was the original "pseudopatient," the first of eight sane, healthy men and women who, almost fifty years ago, voluntarily committed themselves to psychiatric institutions to test firsthand if doctors and staff could distinguish sanity from insanity. They were part of a famously groundbreaking scientific study that, in 1973, would upend the field of psychiatry and fundamentally change the national conversation around mental health. That study, published as "On Being Sane in Insane Places," drastically reshaped psychiatry, and in doing so sparked a debate about not only the proper treatment of the mentally ill but also how we define and deploy the loaded term *mental illness.*

For very different reasons, and in very different ways, "David Lurie" and I held parallel roles. We were ambassadors between the world of the sane and the world of the mentally ill, a bridge to help others understand the divide: what was real, and what was not.

Or so I thought.

In the words of medical historian Edward Shorter, "The history of psychiatry is a minefield." Reader: Beware of shrapnel.

■ ■ ■ ■

PART ONE

■ ■ ■ ■

Much Madness is divinest Sense
To a discerning Eye
Much Sense — the starkest Madness
'Tis the Majority
In this, as all, prevail
Assent — and you are sane
Demur — you're straightway dangerous
And handled with a Chain
                    — Emily Dickinson

1
MIRROR IMAGE

Psychiatry, as a distinct branch of medicine, has come far in its short life span. The field has rejected the shameful practices of the recent past — the lobotomies, forced sterilizations, human warehousing. Today's psychiatrists boast a varied arsenal of effective drugs and have largely dropped the unscientific trappings of psychoanalytic psychobabble, the "schizophrenogenic" or "refrigerator" mothers of yesteryear who had been blamed for triggering insanity in their offspring. Two decades into the twenty-first century, psychiatry now recognizes that serious mental illnesses are legitimate brain disorders.

Despite all these advancements, however, the field lags behind the rest of medicine. Most of our major innovations — better drugs, improved therapies — were in play around the time we first walked on the moon. Though the American Psychiatric Association reassures us that psychiatrists are uniquely qualified to "assess both the mental and physical aspects of psychological problems," they are, like all of medicine, limited by the tools at hand. There are not, as of this writing, any consistent objective measures that can render a definitive psychiatric diagnosis — no blood tests to diagnose depression or brain scans to confirm schizophrenia. Psychiatrists instead rely on observed symptoms combined with patient histories and interviews with family and friends to make a diagnosis. Their organ of study is the "mind," the seat of personality, identity, and selfhood, so it should not be surprising that the study of it is more impenetrable than understanding, say, the biology of skin cancer or the mechanics of heart disease.

"Psychiatry has a tough job. In order to get the answers we need, the truth about what's really going on, we need to understand our most complex organ, the brain," said psychiatrist Dr. Michael Meade. "To understand how this physical organ gives rise to the phenomenon of consciousness, of emotion, of motivation, all the complex functions we humans see as possibly distinguishing us from other animals."

Diseases like the one that set my brain

"on fire" in 2009 are called the great pretenders because they bridge medical worlds: Their symptoms mimic the behaviors of psychiatric illnesses like schizophrenia or bipolar disorder, but these symptoms have known physical causes, such as autoimmune reactions, infections, or some other detectable dysfunction in the body. Doctors use terms like *organic* and *somatic* to describe diseases like mine, whereas psychiatric illnesses are considered *inorganic, psychological, or functional.* The whole system is based on this distinction, on categorizing illness as one or the other, and it dictates how we treat patients up and down the scale.

So what *is* mental illness? The question of how to separate sanity from insanity, of how to even define mental illness, rises above semantics, and above deciding what kind of specialist will care for you or your loved one during a time of intense need. The ability to accurately answer this question shapes everything — from how we medicate, treat, insure, and hospitalize to how we police and whom we choose to imprison. When doctors diagnosed me with an organic illness (as in physical, in the body, *real*) as opposed to a psychiatric one (in the mind, and therefore somehow *less real*), it meant that I'd receive lifesaving treatment instead of being cordoned off from the rest of medicine. This separation would have delayed or even derailed my medical team's efforts to solve the mystery in my brain and would have likely led to my disablement or death. The stakes couldn't be higher, yet, as psychiatrist Anthony David told me, "the lay public would be horrified to realize how flawed and arbitrary a lot of medical diagnosis is."

Indeed, this "flawed and arbitrary" diagnostic system has life-altering ramifications for the one in five adults living in the United States who will experience symptoms of mental illness this year. It even more urgently affects the 4 percent of Americans who contend with serious mental illness,[1] a segment of the population whose lives are often shortened by ten to twenty years. Despite all of our medical progress — of which I'm a direct recipient — the sickest among us are getting sicker.

Even if you are one of the lucky few who

1. Serious mental illness is defined by the National Institute of Mental Health to be "mental, behavioral or emotional disorder . . . resulting in serious functional impairment, which substantially interferes with or limits one or more major life activities."

have never questioned the firing of their synapses, this limitation touches you, too. It shapes how you label your suffering, how you square your eccentricities against the group, how you understand your very self. Psychiatrists, after all, were first known as alienists — a choice term that conveys a sense not only of the doctors' outsider status from the rest of medicine and patients' alienation from themselves, but also of being *the other*. "Insanity haunts the human imagination. It fascinates and frightens all at once. Few are immune to its terrors," wrote sociologist Andrew Scull in his book *Madness in Civilization*. "It challenges our sense of the very limits of what it is to be human." It's undeniable: There is something profoundly upsetting about a person who does not share our reality, even though science shows us that the mental maps we each create of our own worlds are wholly unique. Our brains interpret our surroundings in highly specific ways — your blue may not be my blue. Yet what we fear is the unpredictability of a mentally ill "other." This fear emerges from the sneaking realization that, no matter how sane, healthy, or normal we may believe we are, our reality could be distorted, too.

■ ■ ■ ■

Before I turned twenty-four, all I'd really known of madness was from reading a stolen copy of *Go Ask Alice* in elementary school, or hearing about my stepfather's brother who was diagnosed with schizophrenia, or averting my eyes as I passed a homeless person pawing at imaginary enemies. The closest I got to looking it in the eyes was when, as a tabloid reporter, I'd interviewed in prison a notorious sociopath, whose sharp wit made for great copy. Mental illness was cinematic: the genius mathematician John Nash, played by Russell Crowe in *A Beautiful Mind,* drawing equations on chalkboards, or a sexy borderline à la Angelina Jolie in *Girl, Interrupted.* It seemed almost aspirational, some kind of tortured but sophisticated private club.

And then my illness struck, the autoimmune encephalitis that would devastate me, briefly robbing me of my sanity and changing my life. Sharp fragments of that time stay with me a decade later, slivers from my own memories, my family's stories, or my medical records: the early depression and flu-like symptoms, the psychosis, the inability to walk or talk, the spinal taps, the

26

brain surgery. I remember vividly the imaginary bedbugs, which I believed had taken my apartment hostage; falling apart in the *New York Post* newsroom; nearly jumping out the window of my father's third-floor apartment; the nurses I was convinced were really undercover reporters come to spy on me; the floating eyes that terrified me in the bathroom; the belief that I could age people with my mind. I remember, too, the smug, dispassionate psychiatrist who had treated me in the hospital, calling me an "interesting case" and dosing me with what we would later learn were unnecessary amounts of antipsychotic medications. This was around the time that the medical team began to give up on my case, and the words TRANSFER TO PSYCH started to creep into my medical records.

My family, like many families before them, fought against the tyranny of the mental illness label. My parents were resolute: I was *acting* crazy, sure, but *I* was not crazy. There was a difference. I may have seemed violent, paranoid, and delusional, but I was sick. It wasn't *me.* Something had descended upon me in the same way that the flu or cancer or bad luck does. But when the doctors couldn't immediately find a physical cause, nothing concrete to pinpoint and treat like

27

an infection or tumor, their lens shifted. They moved to a possible diagnosis of bipolar disorder, and then to schizoaffective disorder as my psychosis intensified. Given my symptoms, the psychiatrists' diagnoses made sense. I was hallucinating; I was psychotic; I experienced a cognitive decline. No other tests could explain the sudden change. They saw a bipolar patient. They saw a schizoaffective person. They were wrong. But in nearly any other case, they would have been "right."

Psychiatry is not the lone discipline to wander in such diagnostic haziness. The odds are high that in your lifetime, you will one day suffer from a disease whose causes and treatments are still unknown, or you will face a meaningful medical error that could delay proper treatment, hurt you, or contribute to the cause of your death. The list of illnesses without known cause and cure is long — from Alzheimer's disease to cardiac syndrome X to sudden infant death syndrome. It has been estimated that a third of people who see their general practitioners will suffer from symptoms that have no known cause or are deemed "medically unexplained." We don't really know how everyday drugs like Tylenol work, nor do we really know what exactly happens in the brain during anesthesia, even though 250 million people go under every year.

Look at the role that greed, arrogance, and profit-motivated overprescription played in driving the opiate epidemic — it was common practice to prescribe highly addictive medications for pain until we realized the untold damage and death the drugs caused. Accepted dogma often goes through reappraisals.

Medicine, whether we like to admit it or not, frequently operates more on faith than certainty. We can, in some special cases, prevent diseases with vaccines (smallpox, polio, measles, for example), or with healthy living measures (by purifying our drinking water or quitting smoking) and preemptive scans (as is the case with prostate, breast, and skin cancers), but for the most part we are limited in our ability to actually *cure*.

Despite the shared uncertainties, psychiatry is different from other medicine in crucial aspects: No other discipline can force treatment, nor hold people against their will. No other field contends so regularly with a condition like anosognosia, whereby someone who is sick does not know it, requiring physicians to make difficult decisions about how and when to intervene. Psychiatry makes judgments about people — about our personalities, our beliefs, our morality. It is a mirror held up to the society in which it is practiced. One label applied on your medical record by one doctor could easily send you tumbling off into a whole different hospital with your psychiatric records segregated from the rest of your medical records.

Here was where my story diverged from those of so many other patients. Thanks to many lucky factors that helped set me apart — my age, race, location, socioeconomic situation, generous insurance coverage — doctors pushed for more tests, which led to a spinal tap that revealed the presence of brain-targeting autoantibodies. The doctors were confronted with tangible evidence that disproved their psychiatric diagnosis. My illness was now comfortably *neurological*. I had spinal fluid tests, antibody workups, and academic studies to back me up. Doctors could provide a one-sentence explanation for what happened: My body attacked my brain. And there were solutions that could lead to improvement — even a cure. Hope, clarity, and optimism replaced the vague and distant treatment. No one blamed me or questioned if each symptom was real. They didn't ask about alcohol consumption or stress levels or family relationships.

People no longer implied that the trouble was all in my head.

Mine became a triumphant story of medical progress, thanks to cutting-edge neuroscience. *This girl was crazy; now she is cured.* Medicine stands on a pedestal of stories like these — the father with stage four lung cancer who goes into full remission after targeted therapy; the infant who receives cochlear implants and will never have to know a world without sound; the boy with a rare skin disease who is saved by new skin grown from stem cells. Stories like these lend credence to the belief that medicine follows a linear path of progress, that we are only moving forward — unlocking mysteries of the body and learning more about the final frontiers of our minds on our way to cures for everyone.

I spent four years after my diagnosis collecting facts about my disease, about ages of onset, and about new advances in infusion treatments — a kind of armor to defend against the lonely irrationality of it all. *I am proof of our advancement.* Still, I am stalked by the ever-present threat that psychosis will return. Writing this now, halfway through my pregnancy with twins, I can't forget the ways my body can (and has) failed me. As traumatic as being diagnosed with melanoma was in my late teens, it did not feel like the disease touched *a part of my soul* the same way that my experience with psychosis did. Psychosis is the scariest thing that has ever happened to me. It was neurological, "organic," but it came *from me,* from inside who I am, making it far scarier than any other "physical" illness. It rocked my sense of self, my way of seeing the world, my comfort in my own skin, and shook the foundations of who I am. No amount of fact-gathering could arm me against this truth: We are all hanging on by a very thin thread, and some of us won't survive our fall.

I published *Brain on Fire* to help raise awareness of my condition and in the aftermath was invited to lecture widely at medical schools and neurological conferences, spreading the word about my disease like a missionary, determined to make sure no others were left undiagnosed. At one point, I had the chance to address a large crowd of psychiatrists inside a functioning psychiatric hospital. It was located in a renovated army barracks, but it felt light, white, and modern. *Like a real hospital,* I remember thinking. (When I had packed for the trip I made sure to bring my most adult, sophisticated, *not crazy* ensemble, a simple black-and-turquoise Ann Taylor shift dress paired with a crisp black blazer.)

After my presentation that day, a psychiatrist introduced himself to our group of presenters, speaking in soft but urgent tones about one of his patients. He had diagnosed a young woman with schizophrenia, but in his words, "It just didn't feel right." In fact, she reminded him of me. The woman was of a similar age, had a similar diagnosis, and exhibited similar symptoms. But she also appeared similar to the sea of others with serious mental illness who were being treated alongside her. The question was, How do we know the difference? How to decide who will respond to the intervention I received — the infusions that helped stop my body from fighting itself — versus psychiatric treatments? The group of doctors discussed next steps, the blood tests, lumbar punctures, and MRI scans that might offer an alternative diagnosis for this young woman. Later, as we walked through one of the hospital's units, passing a group therapy meeting, I couldn't help wondering, *Is she in there?*

I learned after my talk that the young woman had indeed tested positive for autoimmune encephalitis, the same disease I'd had. But because she had remained misdiagnosed for two years, unlike the single month I had spent in the hospital, she would probably never regain the cognitive abilities that she had lost. She could no longer care for herself in even the most basic ways and despite her successful diagnosis, she now would, one doctor told me, operate as a permanent child.

I had thought I was done examining my own story after I published my memoir. But once you've come face-to-face with real madness and returned, once you've found yourself to be a bridge between the two worlds, you can never turn your back again. I couldn't shake the thought of the words TRANSFER TO PSYCH in my own medical records. What happened to this young woman almost happened to me. It was like seeing my reflection through the looking glass. She was my could-have-been, my mirror image.

How are we — my mirror images and I — any different from the millions of people with serious mental illness? How could we be so easily misdiagnosed? What does mental illness mean, anyway, and why would one affliction be more "real" than another? These questions have haunted me ever since my memoir was released, when the stories of people's battles within the medical system first landed in my inbox. Some write hoping to have my disease. Anything, some say, except mental illness.

One email I received was from the father of a thirty-six-year-old man who had struggled for two decades with debilitating psychosis. He told me how little modern medicine had been able to offer. "They seem to blame my son for his 'psychiatric illness' on the basis that he has no 'physical illness' that they can heal," he wrote. The drugs, the only treatment offered, had not helped, but actually made him worse. Despite his family's pleas for other options, the response was, "Take the drugs — or we'll force him to take them."

The father recognized his family's plight in my own story and had been inspired by my parents' successful pushback against the medical system. My recovery bolstered his determination to continue searching for more meaningful answers for his son. But something I'd said subsequently had troubled him. In his email he included a You-Tube link to an event where I'd spoken at the release of the memoir's paperback edition. As I watched the clip, I felt like I was being slapped in the face by my own palm. He quoted my words back to me: "My illness appeared as if it was a psychiatric condition, but it was *not* a psychiatric condition — it was a physical condition."

This father felt betrayed hearing me utter the same unfair distinction that he so often heard from his son's doctors. "The brain is a physical organ and physical disease occurs within the brain. Why does that make it a 'psychiatric condition' instead of a physical 'disease'?" he wrote. "What am I missing?"

He was right, of course. How had I so wholeheartedly embraced the same unproven dichotomy that could have confined me to a psychiatric ward, or even killed me? Was it my need to believe that, because I had a physical disorder, I had been "cured" in a way that set me apart from people with psychiatric conditions? What else had I — had *we* — accepted as fact that may have been dangerously reductive? How many fallacies about the mind and brain have we all just been taking for granted? Where did the divide lie between brain illness and mental illness, and why do we try to differentiate between them at all? Have we been looking at mental illness all wrong?

To answer this, I had to heed the advice that my favorite doctor, my own Dr. House, neurologist Dr. Souhel Najjar, often gives his residents: "You have to look backward to see the future."

2
NELLIE BLY

New York, 1887

The young woman fixed her focus on the face, barely registering the wide, mournful eyes that stared back at her in the mirror. She smiled. She raged. She grimaced. She read aloud ghost stories until she spooked herself so thoroughly that she had to turn up the gaslight before she could return to the mirror. She practiced these hideous gazes until dawn, when she cleaned herself, put on an old, moth-eaten dress, and tried to tamp down the growing uncertainty about what lay ahead. There was a chance she would never come home, or that even if she did, this assignment could change her forever. "The strain of playing crazy," she wrote, "might turn my own brain, and I would never get back."

Despite her intense hunger, she skipped breakfast and headed to the Temporary Home for Females on Second Avenue. This morning she called herself Nellie Brown, though she had been born Elizabeth Jane Cochran, and as a professional journalist went by Nellie Bly. Her assignment, given to her by her editor at Joseph Pulitzer's *New York World,* was to infiltrate the notorious Women's Lunatic Asylum on Blackwell Island as a mental patient to write a "plain and unvarnished" first-person narrative about the conditions there. In order to get inside the Blackwell Island asylum, after all, she would need to "prove" that she was indeed insane. This was why she had forced herself to stay up all night, hoping that the physical strain of the sleep deprivation, combined with her disheveled appearance and wild eyes, might induce the house's matron to call the authorities to whisk Nellie off to an asylum, setting the whole plan in motion.

When the US government started tracking the incidence of mental illness, it broke it down into two broad categories of "idiocy" and "insanity." By 1880, the census had expanded to include seven categories of mental disease (mania, melancholia, monomania, paresis, dementia, epilepsy, and dipsomania), but in the first half of the nineteenth century most doctors believed that craziness was one-size-fits-all, something called unitary psychosis. If you acted crazy, you were crazy.

Almost anything could make you a ward of the state. "Compulsive epilepsy, metabolic disorders, syphilis, personality due to epidemic encephalitis, moral adverse conditions such as: loss of friends, business troubles, mental strain, religious excitement, sunstroke, and overheat," read one intake log from California's Patton State Hospital archive. One reason for commitment at Patton State in the nineteenth century was excessive masturbation. Another was for being "kicked in the head by a mule." Other hospital records show that some poor souls were committed for "habitual consumption of peppermint candy" or "excessive use of tobacco." Unmoored after a child died? You could be institutionalized. Use a foul word or two? In a cell you go. Miss a menstrual cycle and you could be committed. These kinds of convenient diagnoses, the sort given to citizens who don't conform, have littered the annals of psychiatry. Hysteria was lobbed at women who dared defy social mores. In England, militant suffragettes, in particular, were diagnosed with "insurgent hysteria." A nineteenth-century Louisiana physician outlined two "conditions" unique to the slaves he studied: dysaesthesia aethiopica, or pathological laziness; and drapetomania, the (evidently inexplicable) desire to escape bondage. Treatments for both included whippings. These were not, in any medical or scientific sense, real illnesses or disorders — they were pseudoscience, purely societal strictures posing as medicine.

Throw a rock into a crowd in the late 1800s, and there's a good chance you'd hit someone who had spent some time in an asylum. And, for those who did end up committed, odds weren't great that they would make it out intact. Once declared insane, you could permanently lose custody of your children, property, and rights to inheritance. Many would remain locked away for a long time, if not the rest of their lives. Those who pushed back often were beaten or "treated" with bleeding, leeching, enemas, and induced bouts of intense vomiting (which were key parts of general medicine's arsenal of care at the time). A substantial portion of people admitted to psychiatric hospitals in this period died within months, even weeks, of being admitted — though there is no definitive proof whether this is because they really suffered from misdiagnosed life-threatening medical conditions or whether the hospitals' conditions themselves led to an early end, or if it was a combination of the two.

The malleability of the era's definitions of insanity meant that any man of a certain means and pedigree could just pay off a doctor or two and dispatch whomever he wanted gone, a disobedient wife, for example, or an inconvenient relative. This understandably bred a widespread anxiety over false diagnoses. Newspapers stoked this fear by publishing a litany of articles about people sidelined into mental hospitals who weren't truly sick.

There was Lady Rosina, an outspoken British writer whose feminist views estranged her from her famous husband, writer Sir Edward Bulwer-Lytton (creator of the most clichéd opening line of all time: "It was a dark and stormy night"). Sir Bulwer-Lytton didn't have time for such a mouthy wife, especially with his seat in Parliament in jeopardy, so he tried to lock her up to shut her up. Thanks to her own celebrity and the pressure that the press put on her husband, she emerged three weeks later and wrote about her experience in 1880's *A Blighted Life.* "Never was a more criminal or despotic Law passed than that which now enables a Husband to lock up his Wife in a Madhouse on the certificate of two medical men, who often in haste, frequently for a bribe, certify to madness where none exists."

Elizabeth Packard continued Lady Rosina's fight in America. Packard butted heads with her Presbyterian minister husband, Theophilus, about her interest in spiritualism. Her religious interests made Packard a direct threat to her husband's stature in the community, so to save his own reputation he recruited a doctor to denounce her as "slightly insane" and commit her to Jacksonville Insane Asylum, where she lived for three years. When Packard was released into her husband's care, she managed to escape the room he had locked her in by dropping a note out of the window. This note reached her friend, who arranged for a group of men to request a writ of habeas corpus on her behalf, giving Packard the opportunity to defend her sanity in court. A jury deliberated for only seven minutes before concluding that, despite what her husband and doctors said, Packard was sane. She published the book *The Prisoners' Hidden Life,* which also featured the experiences of other women unloaded into hospitals by their loved ones. Thanks to her work, the state of Illinois passed a "Bill for the Protection of Personal Liberty," which guaranteed that all who were accused of insanity would be able to defend themselves in front of a jury — since doctors, it was recognized, could be bought and sold. (There were negatives to Packard's reforms, as jurors could be grossly ignorant about matters related to mental illness.)

After Bly successfully made enough of a scene at the boardinghouse for the police to be summoned, she was escorted to Manhattan's Essex Market Police Court, where she faced the judge who would decide whether or not she should be locked up. Lucky for her, or rather for the *New York World,* the judge accepted the events of the morning at face value.

"Poor child," mused Judge Duffy, "she is well dressed, and a lady . . . I would stake everything on her being a good girl." Though she'd worn her most ragged clothes and acted as insane as she could, her genteel looks and manners made it hard for him to take the next step. The judge understood that Blackwell Island was far from a place of refuge, and he hesitated to send someone he felt was too well bred to suffer the indignities there. "I don't know what to do with the poor child," the judge said. "She must be taken care of."

"Send her to the Island," suggested one of

the officers.

The judge called in an "insanity expert," a colloquial term from the era to describe the doctors who chose to work with the insane. These specialists, also called alienists and medical psychologists, or mocked as "bughouse doctors," "quacks," or "mad doctors," mainly spent their careers confined, like their charges, to asylums. (*Psychiatrist* would become the preferred term in the early twentieth century.)

The insanity expert asked Bly to say "ah" so he could see her tongue. He shined a light into her eyes, felt her pulse, and listened to the beating of her heart. Bly held her breath. "I had not the least idea of how the heart of an insane person beat," she later wrote. Apparently her vital signs spoke for her: On whatever quantitative grounds he found to set her apart from the sane, the expert took her to the insane ward at Bellevue. There she was examined by a second doctor who deemed her "positively demented" and shipped her off to Blackwell Island.

When Bly stepped off the boat and onto the shore, the whiskey-soaked attendant welcomed her to the women's asylum: "An insane place, where you'll never get out of."

The word *asylum* comes from an ancient

Greek word meaning "safe from being seized" (by, say, a Homeric warrior). Among the Romans, the word evolved to its current meaning — "a place of refuge" or "a place safe from violence." The first asylums built specifically to house the mentally ill emerged in the Byzantine Empire around AD 500, and by the turn of the new millennium many towns in Europe, the Middle East, and the Mediterranean had one. As forward thinking as that seems, hospitals as we know them today are a modern concept. In the early days, there weren't many differences among jails, poorhouses, and hospitals, and these "asylums" were known for their brutal treatment of their charges.

The vast majority of the mentally ill lived with their families, but this, too, sounds more idyllic than the reality. In eighteenth-century Ireland, mentally ill family members were held in holes five feet beneath their cottage floors, a space not big enough for most to stand up, with a barrier over the hole to deter escape. ("There he generally dies.") The rest of Europe around that time was no more progressive. In Germany, a teenager suffering from some unnamed psychological affliction was chained up in a pigpen for so long that he lost the use of his legs; in England, the mentally ill were staked to the ground in workhouses; in one Swiss city, a fifth of the mentally ill were under constant restraint at home.

Europe's oldest psychiatric hospital, Bethlem Royal Hospital (nicknamed "Bedlam"), started as a priory in London in 1247 and was a hospital in the medieval sense: a charitable institution for the needy. Bethlem began catering exclusively to the insane about a century later; their idea of a cure was to chain people in place and whip and starve them to punish the disease out of their systems. One person, confined to Bethlem for fourteen years, was held by a "stout iron ring" around his neck with a heavy chain that was attached to the wall, allowing him to move only a foot. The belief then was that the insane were no better than animals and should be treated even worse because, unlike livestock, they were useless.

In the mid-1800s, American activist Dorothea Dix deployed her sizable inheritance to devote herself to these issues with a fierceness of purpose that hasn't been matched since. She traveled more than thirty thousand miles across America in three years to reveal the brutalities wrought upon the mentally ill, describing "the saddest picture of human suffering and degradation," a woman tearing off her own skin, a man forced to live in an animal stall, a woman confined to a below-ground cage with no access to light, and people chained in place for *years.* Clearly, the American system hadn't improved much on Europe's old "familial" treatments. Dix, a tireless advocate, called upon the Massachusetts legislature to take on the "sacred cause" of caring for the mentally unwell during a time when women were unwelcome in politics. Her efforts helped found thirty-two new therapeutic asylums on the philosophy of moral treatment. Dorothea Dix died in 1887, the same year that our brave Nellie Bly went undercover on Blackwell Island, in essence continuing Dix's legacy by exposing how little had truly changed.

Blackwell Island was supposed to have been different. Built as a "beacon for all the world," it was located on 147 acres in the middle of the East River and was meant to embody the theory of moral treatment that Dix had championed. Its central tenets came from French physician Philippe Pinel, who is credited with breaking his charges free of their chains (literally) and instating a more humanistic approach to treating madness — though his legacy, historians suggest, comes more from myth than reality. "The mentally sick, far from being guilty people deserving of punishment, are sick people whose miserable state deserves all the consideration that is due to suffering humanity," Pinel said.

Connecticut physician Eli Todd introduced moral treatment Stateside and outlined the new necessities: peace and quiet, healthy diet, and daily routines. These new "retreats" replaced the old "madhouses" or "lunatickhouses" and moved to soothing surroundings away from the stresses of the city. In some cases, asylums expanded into mini-cities, where hospital superintendents, doctors, and nurses lived alongside patients. They tended farms together, cooked in the kitchen together, even made their own furniture and ran their own railroads. The idea was that orderly routines and daily toil created purpose and purpose created meaning, which led to recovery. The doctor-patient relationship was key. People were treated as people, and the sick could be cured.

That was the intention, anyway. Blackwell Island may have been founded on these ideals in 1839, but by Nellie's era it had thoroughly earned its notoriety as one of the deadliest asylums in the country. After Charles Dickens visited in 1842, he immediately wanted off the island and its "lounging, listless, madhouse air." (Dickens later tried to commit his wife, Catherine, to an asylum so that he could pursue an affair with a younger actress — a downright monstrous act considering what he knew of these places.) Blackwell's asylum housed numbers that far exceeded its capacity. In one instance, six women were confined to a room meant for one. Reports detailed "the onward flow of misery," including a woman made to give birth in a solitary cell alone *in a straitjacket,* and another woman who died after mistaking rat poison for pudding.

The inhabitants Bly encountered on Blackwell Island looked lost and hopeless; some walked in circles, talking to themselves; others repeatedly insisted that they were sane but no one listened. Bly, meanwhile, dropped all pretense of insanity once she made it inside the hospital: "I talked and acted just as I do in ordinary life. Yet strange to say, the more sanely I talked and acted, the crazier I was thought to be," she wrote. Any worry — which would soon turn to hope — that she might be exposed as a fake evaporated the minute the nurses plunged her into an ice bath and scrubbed her until her goosefleshed skin turned blue, pouring three buckets of water over her in succession. She was so caught by surprise that she felt she was drowning (a similar sensation, I imagine, to waterboarding). "For once I did look insane," she said. "Unable to control myself at the absurd picture I presented, I burst into roars of laughter."

The first day, she quickly learned what it was like to be discarded by humanity. Whatever ladylike manner had caught the judge's eye was meaningless here, where she was just another in a series of worthless paupers. Patients — even those with open syphilitic sores — were made to wash in the same filthy bathtub until it became thick and dirty enough with human waste and dead vermin that the nurses finally changed it. The food was so rotten that even butter turned rancid. The meat, when offered, was so tough the women chomped down on one end and pulled at the other with both hands to rip it into digestible pieces. Bly had too much decorum to discuss this in her article, but even using the toilets was a traumatic experience. They were long troughs filled with water that were supposed to be drained at regular intervals — but, like everything else on this godforsaken island, what was supposed to happen rarely did.

Bly listened to the stories from her sisters on Hall 6. Louise Schanz, a German immigrant, had landed in this hell simply because she couldn't speak English. "Compare this with a criminal, who is given every chance to prove his innocence. Who would not rather be a murderer and take the chance for life than be declared insane, without hope of escape?" Bly wrote.

Another patient told Bly about a young girl who had been beaten so badly by the nurses for refusing a bath that she died the next morning. One of the "treatments" used on the island was "the crib," a terrifying contraption in which a woman was forced to lie down in a cage so confining that it prevented any movement — like a tomb.

Within a few days, Bly had gathered more than enough evidence for her exposé, but now she began to worry that she would never be free. "A human rat trap," she called it. "It is easy to get in, but once there it is impossible to get out." This was not much of an exaggeration. According to an 1874 report, people spent on average ten to thirty years on Blackwell Island.

By this point, Bly was proclaiming her sanity to anyone who would listen, but the "more I endeavored to assure them of my sanity the more they doubted it."

"What are you doctors here for?" she asked one.

"To take care of the patients and test their

sanity," the doctor replied.

"Try every test on me," she said, "and tell me am I sane or insane?"

But no matter how much she begged to be reevaluated, the answer remained the same: "They would not heed me, for they thought I raved."

Thankfully, after ten days with no word from Bly, her editor sent a lawyer to spring her from the rat trap. Safely back in Manhattan, Bly filed a two-part illustrated exposé — the first called "Behind Asylum Bars," and the second "Inside a Madhouse" — published in the *New York World* in 1887. The article was syndicated across the country, horrifying the public and forcing politicians to do something about it. The Manhattan DA convened a grand jury to investigate and Bly testified, leading jurors on a tour of the island, which had been rapidly scrubbed into shape. But there was only so much Blackwell Island could cover up. In the end, thanks to this young reporter's courage, the Department of Public Charities and Corrections agreed to a nearly 60 percent increase in the annual budget for care of the inmates.

If Bly's publisher hadn't intervened, how long would she have been confined on the island? And what of the other women still trapped inside? The line between sanity and insanity was far less scientific, less quantifiable than anyone wanted to admit. An op-ed in the *New York World* wrote that Bly's exposé showed that "these experts cannot really tell who is and who is not insane," which raised the question of "whether the scientific attainments in mental diagnosis possessed by the doctors who saw her amount to anything or not."

The truth is, at this point in the nineteenth century, alienists still didn't know what to do with the hordes of people filling their asylums. Unsurprisingly, the rest of medicine had no use for these "insanity experts," who seemed to have no expertise in anything. A few years before Bly went undercover, Louis Pasteur had successfully demonstrated the germ theory of disease, leading to the discovery of vaccines against cholera and rabies, which revolutionized medicine by introducing the concept of prevention. In the span of a few decades, medical science had largely dropped the harmful practice of bloodletting and had (decades before Bly's hospitalization) identified leukemia as a blood disorder, helping to launch the new field of pathology. The invisible had at once become visible as medicine bounded into the next century. Yet the alienists, still blind, had only their asylums, cruel "crib" contraptions, and no solid theory about how to explain any of it.

Other than some money being thrown at the problem, nothing changed after Bly's exposé. (As we will see, that would take a much larger mortar shell landing in the heart of psychiatry, nearly a century later.)

One of the most sophisticated and moneyed cities in the world, now aware of such cruelty visited upon its citizens, simply shrugged.

As we still do.

3
THE SEAT OF MADNESS

Today Blackwell Island no longer exists. In 1973, the island was renamed after Franklin D. Roosevelt, and the site where Bly spent her ten harrowing days is now home to a luxury condo development. But the kind of anguish she witnessed there doesn't just disappear. The questions she was trying to answer — questions about what it means to be sane, or insane, what it means to care for a suffering human being who often scares us — remain.

Madness has been dogging humanity for as long as humans have been able to record their own history. But the answer to what *causes* it — where it can be located, in a manner of speaking — has eluded us just as long. The explanation has ping-ponged throughout history among three players: mind/soul, brain, and environment. First, it was believed to be supernatural, a direct effect of meddling by the gods or devils.

56

Thanks to unearthed skulls dated to around 5000 BC, we know that one of the earliest solutions was to bore holes in the head to release the demons that had presumably taken up residency there, a procedure called trephining. Another way to rid oneself of inner demons was to sacrifice a child or an animal so that the evil spirit could trade one soul for another. Early Hindus believed that seizures were the work of Grahi, a god whose name translates quite literally to "she who seizes." The ancient Greeks believed that madness descended on them when their gods were angry or vengeful — a belief that continued on with the teachings of Judaism and Christianity. Lose faith or become too prideful and "the Lord shall smite thee with madness," the Old Testament warned. In the book of Daniel, God punishes Nebuchadnezzar ("those who walk in pride he is able to abase") by deploying a form of madness that transforms him into a raving beast, stripping away his human capacity for rational thought. Exorcisms, ritualistic torture, and even burnings at the stake were some of the approaches employed to release the devil in unquiet minds. Those who survived suicide attempts — seen as an act spurred on by the devil himself — were dragged through the streets and hanged.

Enlightenment thinkers reshaped madness into irrationality and began to think of it as a by-product of the breakdown of reason rather than an outcome of demonic possession. René Descartes argued that the mind/soul was immaterial, inherently rational, and entirely distinct from our material bodies. Though religion clearly still played a role in this thinking, this dichotomy allowed madness to become "unambiguously a legitimate object of philosophical and medical inquiry," wrote Roy Porter in *Madness: A Brief History.*

This area of medical inquiry got a name in 1808: *psychiatrie,* coined by German physician Johann Christian Reil. The new medical specialty (which should attract only the most forward-thinking practitioners, Reil wrote) would treat mind and brain, soul and body — what is today called the holistic approach. "We will never find pure mental, pure chemical, or mechanical diseases. In all of them one can see the whole," Reil wrote. The principles he laid out then are as relevant today: Mental illnesses are universal; we should treat people humanely; and those who practice should be medical doctors, not philosophers or theologians.

Reil's version of psychiatry didn't deter

the many doctors who chased promises of finding the "seat of madness." What causes it? they wondered. Is there one area or hosts of them? Can we be driven to it by circumstance and environment, or is it rooted solely in the organs within our skulls? Alienists began to target the body, expecting that madness could be isolated and targeted — creating some truly horrific treatments along the way, from spinning chairs (developed by Charles Darwin's grandfather Erasmus Darwin) that induced vertigo and extreme vomiting that was believed to lull the patient into a stupor; to "baths of surprise," where floors fell away, dropping people into cold water below to shock the crazy out. As brutal as these treatments were, they were considered a step forward: At least we weren't attributing cause to devils and demons anymore.

An early practitioner named Benjamin Rush, a signer of the Declaration of Independence, believed that the cause of madness was seated in the brain's blood vessels. This prompted him to dream up some deranged treatments, including the "tranquilizing chair" (a case of the worst false advertising ever), a terrifying sensory-deprivation apparatus in which patients were strapped down to a chair with a wooden box placed over their heads to block stimulation, restrict movement, and reduce blood to the brain. Patients were stuck in this chair for so long that the seat was modified to include a large hole that could serve as a toilet. The insane weren't just neglected and ignored; they were abused and tortured — the "otherness" of mental illness making them fair game for acts of outright sadism.

The invention of the microscope led to descriptions of the contours of the brain and nervous system on the cellular level. In 1874, German physician Carl Wernicke pinpointed an area of the brain that, when damaged, created an inability to grasp the meaning of spoken words, a condition called Wernicke's aphasia. In 1901, Frankfurt-based Dr. Alois Alzheimer treated a fifty-one-year-old woman with profound symptoms of psychosis and dementia. When she died in 1906, Alzheimer opened up her skull and found the cause: plaque deposits that looked like tangled-up sections of fibrous string cheese. So: Was her mental illness caused by nothing more than an unfortunate buildup?

The greatest triumph came from the study of syphilis, a disease all but forgotten today (though seeing a resurgence[1]) that surfaced around 1400. The famous people suspected to have had syphilis could crowd a Western civilization Hall of Fame: Vincent van Gogh, Oscar Wilde, Friedrich Nietzsche, Henry VIII, Leo Tolstoy, Scott Joplin, Abraham Lincoln, Ludwig van Beethoven, and Al Capone.

Stories of "the most destructive of all diseases" have abounded since the late Middle Ages. Doctors later called it the "general paralysis of the insane" — a group of doomed patients that made up an estimated 20 percent of all male asylum admissions in the early twentieth century. These patients staggered into hospitals manic and physically off-balance. Some under grand delusions of wealth spent all their money on ridiculous items like fancy hats. Their speech sounded spastic and halting. Over the course of months or years, they would waste away, lose their personalities, memories, and ability to walk and talk, spending their final days sectioned off to the back wards of some local asylum until death. Patient histories, when available, revealed a

1. Syphilis rates are rising across the country. In 2000, there were only 6,000 cases; in 2017, there were 30,644.

pattern: Many of these men and women had developed syphilis sores earlier in their lives. Could this sexually transmitted disease be a latent cause of madness?

The answer came when two researchers identified spiral-shaped bacteria called *Spirochaeta pallida* in the postmortem brains of those of the insane with general paralysis. Apparently, the disease could lie dormant for years, later invading the brain and causing the constellation of symptoms that we now know of as tertiary syphilis. (Syphilis would come to be called the great pox, the infinite malady, the lady's disease, the great imitator, and the great masquerader — one more example of the great pretender diseases, because it could look like a host of other conditions, including insanity.) This was, as contemporary psychologist Chris Frith described, a "kind of peeling of the diagnostic onion." We had parsed out something we thought generally of as "insanity" as having a physical cause. And the best part was that we could eventually cure it if we caught it early enough, too.

(Though they have different causes, the symptoms of syphilis share many similarities with those of autoimmune encephalitis, the disease that struck me, which I guess could give autoimmune encephalitis the dubious honor of being the syphilis of my generation.)

The more we learned about the science of the mind, the hazier the boundary between neurology and psychiatry became. During the twentieth century, neurology broke off into a distinct branch of medicine, and in doing so "claimed exclusive dominion over the organic diseases of the nervous system" — like stroke, multiple sclerosis, and Parkinson's. Meanwhile, psychiatrists took on the ones "that could not be satisfactorily specified by laboratory science" — like schizophrenia, depression, and anxiety disorders. Once a biological breakthrough was achieved, the illness moved out of psychiatry and into the rest of medicine. Neurologists work to uncover how damage to the brain impairs physical function; psychiatrists are there to understand how this organ gives rise to emotion, motivation, and the self. Though the two fields overlap considerably, the separation embodies our mind/body dualism — and this continues today.

Clearly, syphilis and Alzheimer's disease weren't the only causes of insanity. In order to track down and cure the others — if they could be found — psychiatrists still needed to develop a diagnostic language that could help pinpoint the different types (which would hopefully lead to the cleaving out of different causes) of mental illness.

German psychiatrist Emil Kraepelin had been tackling this issue since the late nineteenth century, and though you've likely never heard of him, his work has had more influence on the way psychiatry is practiced today than did the famous Sigmund Freud, born the same year: 1856. The son of a vagabond actor / opera singer / storyteller, Kraepelin dedicated his life to organizing mental illnesses into orderly parts, perhaps as a reaction to such an unorthodox father. In doing so he endowed the nascent field with a new nosology, or system of diagnosis, that would later inspire the *Diagnostic and Statistical Manual of Mental Disorders,* the bible of psychiatry today. Kraepelin studied thousands of cases and subdivided them, breaking down what was described as "madness" into clear categories with varied symptoms as best he could. This culminated in the description of the medical term *dementia praecox.* Kraepelin defined *dementia praecox* in his 1893 textbook *Psychiatrie* as an early onset permanent dementia, a biological illness that caused psychosis and had a deteriorating course with little hope to improve, causing "incurable and permanent disability." Kraepelin separated dementia praecox patients from those with "manic-depressive psychosis," a disorder of mood and emotion that ranged from depression to mania, which had a better long-term prognosis. This division continues today with schizophrenia (and its component parts) and bipolar disorder (and its component parts). (In 1908, almost two decades after Kraepelin presented the diagnosis dementia praecox to the public, Swiss psychiatrist Paul Eugen Bleuler tested out the new term *schizophrenia,* which translates to "splitting of the mind," contributing to a long-running confusion[2] over the term. Later, psychiatrist Kurt Schneider further defined schizophrenia with a list of "first rank symptoms" that include auditory hallucinations, delusions, and thought broadcasting.)

Now, finally, psychiatrists could make predictions about course and outcome. Most important, they could provide a name for their patients' suffering, something I

2. *Schizophrenia* remains one of the most misused medical terms. Enter "schizophrenic" into a Google News search and you'll land on an array of uses describing everything from Brad Pitt's movie *War Machine* to Facebook's new community guidelines — all glaringly incorrect usages.

personally would argue is one of the most important things a doctor can do, even if a cure isn't in sight. Still, the cause remained elusive — as it continues to.

Doctors began to slice and dice their way through "insane" brains. They removed living people's thyroids, women's ovaries, and men's seminal vesicles based on half-baked theories about the genetic origins of madness. An American psychiatrist named Henry Cotton, superintendent of Trenton State Hospital in New Jersey, offered a "focal infection theory" of mental illness, which posited that the toxic by-product of bacterial infections had migrated to the brain, causing insanity. It wasn't a terrible idea in theory (there are infectious causes of psychosis), but Cotton's solutions were a nightmare. In an attempt to eliminate the infection, he began by pulling teeth. When that didn't work, he refused to reconsider and instead removed tonsils, colons, and spleens, which often resulted in permanent disablement or death — and got away with it because his patient population had neither the resources nor the social currency to stop him.

Clinicians and researchers also embraced the growing eugenics movement that argued that insanity was a heritable condition passed down through inferior genes. In America, thirty-two states passed forced sterilization laws between 1907 and 1937 — why not stop the spread of undesirables, they thought, by cutting off their ability to reproduce? The Nazis adopted America's science-approved sadism, sterilizing three hundred thousand or so German psychiatric patients (the most common diagnosis was "feeblemindedness," followed by schizophrenia and epilepsy) between 1934 and 1939 before they took it one step further and began exterminating "worthless lives" — executing over two hundred thousand mentally ill people in Germany by the end of World War II.

In the aftermath of the war, as the full horror of Nazi atrocities hit the American public, the timing seemed overdue for a reassessment of psychiatry and its obsession with finding biological causes for mental illness — especially in 1955, when over a half million people lived in psychiatric hospitals, the highest number ever.

In a strange confluence of events, the same year that Kraepelin popularized dementia praecox, Freud emerged with a new theory of treating the mind called psychoanalysis. While asylum psychiatrists interrogated the body, another group of doctors, psychoanalysts, had moved so far away from the search for an answer in the physical that it was as if they were practicing a different discipline altogether. Psychiatry outside the asylum had little in common with that practiced inside. Outside the asylum, the idea reigned that the *mind* was the seat of all mental suffering, not the gray matter of the brain. For someone like me, so accustomed to talk of neurotransmitters, dopaminergic pathways, and NMDA receptors, the popular terms of that era, like *penis envy, phallic stage,* and *Oedipal conflict,* feel awkward and clumsy, holdovers from a quainter world. But it wasn't that long ago when these were the norms. Every Baby Boomer alive today was born when terms like these dominated the field.

Psychoanalysis invaded the US by way of Europe right before World War II, offering up a new theory that provided fresh insight into mental anguish — and, for once, real cures — as war-weary soldiers returned from battle healthy by all physical estimations, but emotionally unable to join the workforce or engage in family life. For the first time ever, there were more recorded casualties related to the mind than to the body. It was a sobering thought: If a healthy young man could be reduced to a shaking, fearful, hysterical one without any physical cause, then couldn't this happen to any of us?

Freud (who died before psychoanalysis really took off in America) gave us a path out of this dark forest of uncertainty. In his explanation, our minds were divided into three parts: the id (the unconscious — rife with repression and unfulfilled desires); the ego (the self); and the superego (the conscience), all engaged in battle. The analyst's goal was to "make the unconscious conscious" and with a surgeon's focus zero in on the underlying conflict — our libidos, repressed desires, death drives, projections, and wish fulfillment fantasies; all that deep, dark, murky stuff from our childhoods — on the way to insight. There was "nothing arbitrary or haphazard or accidental or meaningless in anything we do," wrote Janet Malcolm in *Psychoanalysis: The Impossible Profession*.

And who wouldn't want this kind of careful attention and promise of a cure over the dour inevitability that the biological side (à la Emil Kraepelin) was offering? Consider the two differing interpretations of a patient's story as analyzed by both Kraepelin's followers and Freud. In 1893, fifty-one-year-old German judge Daniel Paul Schreber started to become obsessed with the idea that to save the world, he needed to become a woman and give birth to a new human race. He blamed these disturbing thoughts on his psychiatrist, whom he called a "soul murderer" who had implanted these delusions via "divine rays." Doctors diagnosed Schreber with Kraepelin's dementia praecox and committed him to a psychiatric hospital, where he eventually died. When Freud read Judge Schreber's account, *Memoirs of My Nervous Illness,* he suggested that, instead, Schreber's behaviors stemmed from repressed homosexual impulses, not from an incurable brain disease. Treat the underlying conflict and you'd treat the person. If you had your choice, which kind of treatment would you pick? Americans overwhelmingly chose Freud, and Kraepelin and his acolytes were forsaken to the professional boondocks.

By the 1970s, nearly every tenured professor in psychiatry was required to train as an analyst, and most textbooks were written by them, too. Overnight, it seemed, analysts got "a power, a secular power, that they never had before and they never had since," psychiatrist Allen Frances told me. You no longer went to your priest or parents; you paid an analyst to shrink you. Now "mind doctors" wanted to mine your "family relations, cultural traditions, work patterns, gender relations, child care, and sexual desire." Psychiatrists were thrilled to leave the back wards of mental hospitals, where difficult patients had few options for cures, and instead to retrain as analysts and cater lucrative talk therapy treatments (five days a week!) to help the so-called worried well who suffered from a case of nerves brought on by modern life. The people who needed help the most were left behind as analysts comfortably cherry-picked their patients — mostly wealthy, white, and not very sick.

Americans jumped on the couch, embracing the "blank screens" of their therapists and the idea that the mind could be improved. Decades after his death, Freud's method was suddenly everywhere: in women's magazines, in advertising (Freud's nephew Edward Bernays is called the father of public relations); even the CIA started snatching up analysts. America's second-biggest bestseller after the Bible became Dr. Benjamin Spock's *The Common Sense Book of Baby and Child Care,* which was based on Freudian theories. Another huge book of the moment was Norman O. Brown's *Life Against Death: The Psychoanalytic Meaning of History,* which attempted to reframe the past through a Freudian battle between freedom and repression. Hollywood hired psychiatrists on retainer on movie sets. Insurance companies paid for months of talk therapy and reimbursed at levels equal to other serious medical procedures.

No matter how many psychiatrists enlisted, however, there still weren't enough. By 1970, despite the influx of doctors, the demand exceeded the supply. Unlike the custodians of the sick in the past, psychoanalysts now promised to listen to their patients. In the best cases, patients found clarity and meaning from this relationship. Instead of pathologizing people outright, analysts saw each patient as unique in her psychic suffering. They gave us a deeper understanding of how fraught and layered our interior lives are: the complexities of sexuality; the key role that our childhoods play in our adult lives; how the unconscious speaks to us through our behaviors. Through the "interchange of words between patient and physician," as Freud put it, you could explore, comprehend, and even heal the sick parts inside us. "Words were originally magic, and the word retains much of its old magical powers even today," Freud wrote in 1920. "Therefore let us not underestimate the use of words in psychotherapy."[3]

One of the varied downsides was that doctors enacted vivid blame games on their patients (and the families of their patients), especially on mothers. (See the *refrigerator mother* [lack of maternal warmth] and the *schizophrenogenic mother* [an overbearing, nagging, domineering female, usually paired with a weak father], both of whom were believed to create symptoms of schizophrenia and autism in their children.) Viennese psychoanalyst Bruno Bettelheim,[4] "psychoanalyst of vast impact," in *The Empty Fortress* in 1967 compared the family structure of those with mental illness, especially autism, to concentration camps, a particularly damning argument because Bettelheim himself had survived two years in Dachau

3. A quick distinction: *Psychotherapy* is a more general term, one interchangeable with talk therapy (though distinct from counseling, which tends to focus on a specific issue), whereas *psychoanalysis* started with Freud and is "the most complex of the talking treatments," according to the British Psychoanalytic Council.

4. I should add that after his suicide in 1990, allegations emerged that Bettelheim exaggerated his credentials, fabricated research, and abused children under his care.

and Buchenwald. The only way one could recover was to completely sever relationships with family.

But what you didn't get with Freud was a focus on diagnosis. In fact, his followers practiced "extreme diagnostic nihilism." Nomenclature, shared diagnostic language — these didn't really matter to the analysts. In fact, psychiatrists expanded the scope of social deviance, pathologizing almost everyone in the process, effectively closing the chasm between sanity and insanity by showing that "true mental health was an illusion," as anthropologist Tanya Marie Luhrmann wrote in her study of the profession *Of Two Minds.* According to a now infamous 1962 Midtown Manhattan study based on two-hour interviews with sixteen hundred people in the heart of the city, only 5 percent of the population were deemed mentally "well." The whole world was suddenly crazy, and psychiatrists were their caped crusaders.

America was again starting to look a lot like it had in the time of Nellie Bly — where anyone could be and often was (mis)diagnosed.

And then, in February 1969, "David Lurie" walked into the intake room at an unspecified hospital in Pennsylvania and set off a metaphorical bomb. He finally proved what so many people had long suspected: Psychiatry had too much power and didn't know what the hell to do with it.

4

On Being Sane in Insane Places

I often imagine Bly's trip back to Manhattan aboard the transport ferry from Blackwell Island — the air whipping her hair, the foul smells of the river, the buzzy relief — as her thoughts turned to the women she had abandoned.

"For ten days I had been one of them. Their sorrows were mine, mine were theirs, and it seemed intensely selfish to accept freedom while they were in bondage," Bly wrote. "I left them in their living grave, their hell on earth — and once again I was a free girl."

That was exactly how I felt every time I thought about my mirror image, and all those who had not been saved as I had — the others whom psychiatry had left behind.

A month or two after my presentation at the psychiatric hospital, I had dinner with Dr. Deborah Levy, a McLean Hospital psychologist who studies (among other things) genes that appear to put people at risk for developing serious mental illness, and her colleague Dr. Joseph Coyle, a McLean Hospital psychiatrist who is one of the foremost experts on the NMDA receptor, a part of the brain that is tampered with in the illness that struck me. (Tracking two neuroscience researchers in conversation is much like following an intense hockey game. Take your eye off the puck for one second, and you're lost.) We spoke about the hysterias of the past and the conversion disorders of the present; about the difference between malingering and Munchausen syndrome. The former describes faking an illness for some kind of gain (to win a lawsuit, for example), while the latter is the name of a mental disorder in which one pretends to be sick when there isn't any obvious incentive. (The famous case of Gypsy Rose Blanchard is an extreme example of Munchausen *by proxy,* when you make someone else sick, often a child.) We talked a bit about the great pretender illnesses that blur the boundary between psychiatry and neurology and how hard it is for physicians to parse those out and about how my disease appeared to be a bridge between the two worlds, a "physical" disorder that masked itself as a "psychiatric" one.

I chimed in with the story I had recently learned of my mirror image. There shouldn't have been any difference between us; she should have received the same treatment, she should have had the same quick and urgent interventions, and she should have had the opportunity to recover as I had. But she had been derailed because of one crucial difference: Her mental diagnosis had stuck. Mine hadn't. Sympathetic, Dr. Levy asked me if I had ever heard of the study by Stanford professor David Rosenhan.

"Do you know it? The one where the people purposefully faked hearing voices and were admitted to psychiatric hospitals and diagnosed with schizophrenia?" she asked.

Nearly fifty years after its publication, Rosenhan's study remains one of the most reprinted and cited papers in psychiatric history (despite being the work of a psychologist rather than a psychiatrist). In January 1973, the distinguished journal *Science* published a nine-page article called "On Being Sane in Insane Places," whose driving thesis was, essentially, that psychiatry had no reliable way to tell the sane from the insane. "The facts of the matter are that we have known for a long time that diagnoses are often not useful or reliable, but we have nevertheless continued to use them. We now know that we cannot distinguish insanity from sanity." Rosenhan's dramatic conclusions, backed up for the first time by detailed, empirical data and published by *Science,* the sine qua non of scientific journals, were "like a sword plunged into the heart of psychiatry," as an article in the *Journal of Nervous and Mental Diseases* observed three decades later.

Rosenhan, a professor of both psychology and law, had posed this opening salvo: "If sanity and insanity exist, how shall we know them?" Psychiatry, it turned out, didn't have an answer — as it hadn't for centuries. This study "essentially eviscerated any vestige of legitimacy to psychiatric diagnosis," said Jeffrey A. Lieberman, chairman of Columbia's Department of Psychiatry. In the wake of the study's publication, "Psychiatrists looked like unreliable and antiquated quacks unfit to join in the research revolution," added psychiatrist Allen Frances.

By the late 1980s, a little over a decade after its publication, nearly 80 percent of all intro-to-psychology textbooks included Rosenhan's study. Most histories of psychiatry devote at least a section to it — even in the pocket-size *Psychiatry: A Very Short Intro-*

duction (a kind of "psychiatry for dummies"), which is only 133 pages long, the Rosenhan study takes up nearly a whole page on "psychiatric gullibility." To this day, "On Being Sane in Insane Places" is taught in a majority of psych 101 classes, an outright coup for a study four decades old. Its power was in its scientific certainty. Journalists, writers, even psychiatrists had infiltrated the world of the mentally ill before Rosenhan and exposed the horrors there — but none had done so with such rigor, with such a broad sample set, with such extensive citations, in such an attention-grabbing way, at just the right time in just the right publication. These researchers were not "a bunch of harum-scarum sensationalists," one newspaper reporter wrote, but a varied group gathered by Rosenhan, a highly credentialed man who boasted dual professorship in law and psychology at Stanford University. Rosenhan's study, published in one of the world's most prestigious academic journals, quantified the medications, the number of minutes per day the staff spent with the patients, even the quality of those interactions. Unlike Nellie Bly and others before and after, David Rosenhan's data was, at last, unimpeachable.

Eight people — Rosenhan himself and seven others, a varied group that included three women, five men, a graduate student, three psychologists, two doctors, a painter, and a housewife — volunteered to go undercover in twelve institutions in five states on the East and West Coasts and present with the same limited symptoms: They would tell the doctors that they heard voices that said, "thud, empty, hollow." (One potential pseudopatient who had not conformed to Rosenhan's rigorous data collection methods was, as explained in a footnote, pulled from the study.) With this standardized structure, the study tested whether or not the institutions admitted the otherwise sane individuals. Based on those symptoms alone, the psychiatric institutions diagnosed all the "pseudopatients" with serious mental illnesses — schizophrenia in all cases but one, in which the diagnosis was manic depression. The length of hospitalization ranged from seven to fifty-two days, with an average of nineteen days. During their hospitalizations, twenty-one hundred pills — serious psychopharmaceuticals — were prescribed and administered to these healthy individuals. (The pseudopatients were trained to "cheek" or pocket the pills so they could be spit out in the toilet or thrown away rather than ingested.)

Beyond a few biographical adjustments for privacy reasons, the pseudopatients used their own life stories. Once inside their designated institution, it was up to them to get themselves out. "Each was told that he would have to get out by his own devices, essentially by convincing the staff that he was sane," Rosenhan wrote. Just as Nellie Bly had done nearly a century earlier, they dropped their hallucinations as soon as they were admitted and behaved "normally," or as normally as the bizarre conditions allowed. Yet, from the moment of admittance, clinicians viewed all behaviors through the prism of the pseudopatients' presumed mental illness. No pseudopatient was unmasked by the staff, yet 30 percent of fellow patients in the first three hospitalizations noticed something was awry, commenting, in one case, "You're not crazy. You're a journalist or a professor. You're checking up on the hospital." Nurses' reports noted that "patient engages in writing behavior" when the pseudopatient was observed calmly documenting the activities of the ward for his or her undercover research. "Having once been labeled schizophrenic, there is nothing the pseudopatient can do to overcome the tag. The tag profoundly colors others' perceptions of him and his behavior," Rosenhan wrote.

"How many people, one wonders, are sane but not recognized as such in our psychiatric institutions?" asked Rosenhan. "How many patients might be 'sane' outside the psychiatric hospital but seem insane in it — not because craziness resides in them, as it were, but because they are responding to a bizarre setting?" Or, as the nurse's comment about "writing behavior" revealed, simply exhibiting normal behaviors that are misinterpreted as abnormal under the label of mental illness. It was unusual for a paper as narrative as this to end up in *Science,* one of the most widely read peer-reviewed academic journals in the world, endowed with seed money from Thomas Edison and later Alexander Graham Bell. (*Science*'s most famous papers include the first time the entire human genome was sequenced, early descriptions of the AIDS virus, a paper on gravitational lensing by Albert Einstein, and one on spiral nebulae by astronomer Edwin Hubble.) That it was published in such a revered general science academic journal gave the study a life that no one — probably not even David Rosenhan himself — could have seen coming.

Arriving on the scene when it did, Rosenhan's "On Being Sane in Insane Places" ended up falling right in line with other, more theoretical rebukes that had been building from inside the ranks of psychiatry from people who asserted that mental illness didn't even exist. The pendulum had swung once more, this time into a third position, moving from the idea that mental illness resided in the brain as a tangible disease, like cancer, to the theory that it emerged from unresolved conflict in the mind's psyche, to the new conviction that the "illness" itself lay entirely in the eye of the beholder. Intentionally or not, Rosenhan's study ultimately built on this idea, arguing that the healthy volunteers were deemed insane *because* they were in an insane asylum, not because of any objective, external truths that psychiatry could point to for a diagnosis. Rosenhan provided the key element missing from anti-psychiatry's arguments — proof of its convictions.

The timing of the study couldn't have been more fraught for psychiatry. These were the early rumblings of psychiatry's worry years. Sobering studies cast psychiatry in a less-than-effective light. In 1971, a large-scale US/UK study showed that there was little consensus across the Pond about schizophrenia. American psychiatrists worked with a broader concept of the disorder and overwhelmingly diagnosed people with it, while British doctors were more likely to diagnose patients with manic depression, now known as bipolar disorder. Two psychiatrists on the same side of the Atlantic, studies showed, agreed on diagnosis less than 50 percent of the time — worse than blackjack odds. American psychiatrist Aaron T. Beck, who would later father the field of cognitive behavioral therapy, published two pieces on the lack of reliability in psychiatric diagnosis, concluding in his 1962 paper that psychiatrists agreed only 54 percent of the time when diagnosing the same psychiatric patient.

Meanwhile, psychiatric hospitals closed at a rapid clip across the country. By the time California governor Ronald Reagan took office in 1967, state hospitals had released half of all their patient population. Under Reagan's leadership, California passed several acts that hastened the demise of the institutions across the state — and the rest of the country followed. Yet even as the hospitals were being closed, psychiatry's reach was spreading wide outside the asylum, like ground ivy, into Hollywood, government, education, child-rearing, politics, and big business, enjoying a sudden social cachet while turning its back on the people who needed help the most — the seriously mentally ill.

Society at large, it seemed, was ready to push back against this overextension. In the wake of his study, David Rosenhan became an academic celebrity, a media darling whose research was extensively covered in the nationwide press. It launched scores of articles, some of them outright hostile, everywhere from the *New York Times* to the *Journal of Abnormal Psychology,* as people debated the limits of psychiatry as a medical specialty. (Various Reddit pages dedicated to the study still spring up with thousands of commenters weighing in, embracing the idea that there exists a respected academic paper they can brandish to jab back at a medical specialty that, to their minds, has ignored, exploited, or abused them.) There was even a rash of pseudopatient copycats in the 1970s — including one college student at Jacksonville State Hospital who was unmasked as a faker by the staff in 1973. He was the second pseudopatient outed there in a period of six months.

The study brought Rosenhan renown as a respected expert in diagnosis, precisely because of his critique of it. (This happened despite the fact that he had spent only six months in a hospital setting early in his career, when he researched — but never treated — people with serious mental illness.) He testified in a Navy hearing about the schizophrenia diagnosis and involuntary commitment of a skipper, worked as a psychology consultant to the Veterans Administration, and became a mascot for the limitations of psychiatry at countless academic conferences. Lawyers cited Rosenhan's study as proof that a psychiatrist as an expert witness was an oxymoron — claiming that in the courtroom such testimony was as legitimate as "flipping coins."

When Dr. Deborah Levy introduced me to the study, I didn't yet know how the tentacles of this one almost fifty-year-old paper extended in so many wild directions that it was cited to further movements as disparate as the biocentric model of mental illness, deinstitutionalization, anti-psychiatry, and the push for mental health patient rights. Nor did I know that it would alter my perspective on something that I thought I had all figured out. Reading the study for the first time, I — like many before me — simply recognized so much of my own experience in Rosenhan's words. I had seen how doctors' labels altered the way they saw me: During my hospitalization, one psychiatrist described my plain white shirt and black leggings as "revealing," for instance, and used it as proof that I was hypersexual, a symptom that supported her bipolar diagnosis. It's hard to ignore the judgment that comes with those kinds of labels. Yet the minute the doctors discovered my issues were neurological — after I had spent weeks living with a psychiatric diagnosis — the quality of care improved. Sympathy and understanding replaced the largely distant attitude that had defined my treatment, as if a mental illness were my fault, whereas a physical illness was something unearned, something "real." It was the same way the psychiatrists treated the pseudopatients when the cause of their presumed distress could only be "mental."

"It is not known why powerful impressions of personality traits, such as 'crazy' or 'insane,' arise," Rosenhan wrote. "A broken leg is something one recovers from, but mental illness allegedly endures forever. A broken leg does not threaten the observer, but a crazy schizophrenic? There is by now a host of evidence that attitudes toward the mentally ill are characterized by fear, hostility, aloofness, suspicion, and dread. The mentally ill are society's lepers."

I identified with the extreme loss of self that all eight pseudopatients experienced during their hospitalizations — and bristled at the blame directed at the pseudopatients, as if they didn't deserve sympathy or care. "At times, depersonalization reached such proportions that pseudopatients had the sense that they were invisible, or at least unworthy of account," Rosenhan wrote. I recognized their outrage over the blatant hubris of the doctors who in the face of uncertainty doubled down with an unquestionable infallibility. "Rather than acknowledge that we are just embarking on understanding, we continue to label patients 'schizophrenic,' 'manic-depressive,' and 'insane,' as if in those words we had captured the essence of understanding. The facts of the matter are that . . . we cannot distinguish insanity from sanity," Rosenhan wrote.

In my first reading of "On Being Sane in Insane Places," in a quiet Boston hotel room, the first of hundreds of readings to come, I saw immediately why so much of the general public had hailed it — and why psychiatry writ large despised it. I recognized the validation Rosenhan's work gave

89

to that father who had emailed me. I pin-pointed so much of my own disappointment and frustration as a former patient myself. And I could feel, viscerally, the undercurrent of rage that travels through his paper that I feel, too, when I picture the face of my mirror image, that anonymous young woman, trapped in a psychiatric diagnosis, who would never be the same.

"You are a modern-day pseudopatient," Dr. Levy said to me over our dinner that night, meaning that I was also misidentified as a psychiatric patient.

I took it a different way: It was a challenge, a call to learn more and understand how this study, and the dramatic questions Rosenhan raised almost fifty years ago, could help the untold others whom our health care system still leaves behind.

5
A RIDDLE WRAPPED IN A MYSTERY INSIDE AN ENIGMA

I had so many questions for David Rosenhan: about his experiences, about the pseudopatients, about the creation of and the challenges in implementing the study. But he had died in 2012, in those same months when I was preparing for the release of my own *Brain on Fire.* I searched eagerly for more of his work, but with the exception of one companion piece, where Rosenhan clarified some of the points made in his original study, and a short personal reference to the study in an introduction to his abnormal psychology textbook, he never again published on the topic. He had even secured a book deal, I learned, but ended up never delivering the manuscript and was later sued by the publisher for it. He had walked away from this subject that so desperately needed a champion. What had happened to silence him?

Unfortunately, I would not learn the

answer easily. Google searches and basic digging led me nowhere in understanding more about the creation of "On Being Sane in Insane Places." A news clip search revealed no further details. It seemed there was little else to find beyond the original premise — eight anonymous pseudopatients, twelve hospitals, "thud, empty, hollow." None of the pseudopatients had gone public, their names never released. Nor had anyone revealed the identities of the hospitals they infiltrated. Rosenhan had remained tight-lipped his whole life about the identities of the hospitals (with one exception — he did reassure the superintendent of Delaware State Hospital that despite rumors, he had *not* sent pseudopatients there). He was determined to protect their privacy, he wrote, because he didn't blame the individual doctors and hospitals themselves so much as the system overall. Given how groundbreaking the study was, it was startling that such a large part of it remained a mystery almost five decades later.

Secrecy or no, the study had clearly touched a nerve, and not in the same way it had for me. In the April issue of *Science,* following the January publication of "On Being Sane in Insane Places," furious letters to the editor filled twelve whole pages.

"Through the publicity attracted by his methods," one Yale psychiatrist wrote to *Science,* "Rosenhan may have provided society with one more excuse for pursuing the current trend of vilifying psychiatric treatment and neglecting its potential beneficiaries." Another wrote: "It can only be productive of unwarranted fear and mistrust in those who need psychiatric help, and make the work of those who are trying to deliver and teach about quality care that much harder." They were, understandably, standing their ground — but that same ground was now shifting beneath them.

The debate Rosenhan had touched off continued to rage for decades. In 2004, author and psychologist Lauren Slater claimed that she replicated the study. Her work prompted a round of scorching critiques from many of the same members of the psychiatric community who had ripped Rosenhan's study more than thirty years earlier. I marveled at how psychiatry could be so defensive, when so many others had acknowledged the problems before Rosenhan arrived to document them with hard data. Why attack the messenger?

Finally, I stumbled upon a link that got me a little closer to that messenger: A BBC radio report that aired before David Rosenhan's death revealed that Rosenhan's personal files were with his close friend and colleague Lee Ross, a seminal Stanford social psychologist. I soon found myself in a rental car, hopelessly lost on my way to Stanford University's Department of Psychology in Jordan Hall.

"I'm so sorry I am late," I hear myself say to Lee Ross on the audio recording I made of our meeting. I can hear in my voice how painfully aware I am of the stature of the man I'm interviewing. Lee Ross has written well over a hundred research papers, authored three and edited five influential academic books (when I visited him, he was in the midst of co-writing *The Wisest One in the Room,* a book that pushes readers to apply the best of social psychology research to their own lives), and founded the Stanford Center on International Conflict and Negotiation with, among others, psychologist Amos Tversky (a subject of Michael Lewis's recent *The Undoing Project*).

Lee also coined the term *fundamental attribution error,* which theorizes that people are more likely to credit other people's faults to internal factors (*she's late because she's a directionally challenged idiot with no time management skills*) but credit external factors when we think of ourselves (*I'm late because Stanford's campus is needlessly confusing, and it's impossible to find a parking spot*). His research interests range from shortcomings in intuitive judgment and decision making, to sources of interpersonal and intergroup misunderstandings, to "naive realism" — a way of viewing the world that refuses to acknowledge that everyone experiences realities differently. He documented the shortcomings of the "intuitive psychologist" in one of his early papers, which showed how researchers' biases color the interpretations of their data. He studied belief perseverance, or the tendency of people to dig in when presented evidence contrary to their convictions. He also coined the term *false consensus effect* to describe how people often overestimate how common their beliefs are — particularly dangerous in those who hold extremist views.

In other words, if I had to narrow Lee's interest down to a few words, it would be *the fallibility of belief.* And he was close friends with David Rosenhan, the man whose past I had come to mine.

Lee Ross is a kind man but, according to a colleague, "doesn't suffer fools." He speaks slowly. His wildly engaging eyes, his gentle voice, and his congenial way of angling his head in your direction as you try to make a point, seeming to peer right inside you, made me nervous.

When, in my rambling way, I told Lee about how my own story led to David Rosenhan's, he interrupted me.

"I had Guillain-Barré," he said. "I had hallucinations, too. But I had hallucinations because I was severely sleep-deprived, because I couldn't close my eyes. They like to say everybody's about six degrees Fahrenheit from hallucinating."

(Auditory hallucinations, the symptom most associated with serious mental illness, are actually quite common in the general population — as widespread as left-handedness, some studies say. A host of medical conditions can induce them: high fevers, of course, but also hearing loss, epilepsy, alcohol withdrawal, bereavement, and intense stress. If you do hear voices, you're joining an esteemed group that includes Socrates, Sigmund Freud, Joan of Arc, Martin Luther King Jr., and Winston Churchill.)

Guillain-Barré syndrome is an autoimmune disease that occurs when the body's immune system targets nerves, which can sometimes result in paralysis. Lee's case struck him five years before our meeting, and at one point, he could not swallow or talk. It is hard to imagine a worse fate for a man so interested in conversing with the world. After several months of treatment, hooked up to a respirator and a feeding tube, Lee recovered and the lingering effects are minor, if there are any.

Coincidentally, David Rosenhan had suffered from Guillain-Barré, too. Lee mentioned this as he pointed out the office down the hall where Rosenhan had worked for more than thirty years. That two people who shared the same floor of a small office building had had the same rare autoimmune disease shocked one doctor with whom I shared this information — it's a one-in-a-billion chance, the doctor said. But it was true: I would later confirm this coincidence with Rosenhan's family and friends. It was the first of many small, improbable details I would encounter in my investigation.

Before my visit, Lee had set aside a stack of books that had once belonged to Rosenhan and that Lee believed were key to his thinking: *The Myth of Mental Illness* by Thomas Szasz, *Self and Others* by R. D. Laing, and *Asylums* by Erving Goffman — all works associated with the anti-psychiatry movement.

As I thumbed through Rosenhan's books, Lee told me the origin story of their friendship. They had met in the early 1970s when Rosenhan joined Stanford's psychology faculty after leaving Swarthmore College. Stanford in those days was home to an all-star roster of psychologists, including Philip Zimbardo, who led the much-publicized Stanford Prison Experiment in 1971. The observational study, which recently spawned a movie, purportedly simulated prison life in the basement of the university's Jordan Hall with volunteers playing the parts of fake guards and fake prisoners. After a few days, the guards, drunk on their own power, abused the prisoners, who withdrew and grew resigned to their fate. Zimbardo's study was published in 1973, not long after Rosenhan's. The Stanford Prison Experiment made Zimbardo a legend the same way "On Being Sane in Insane Places" did for Rosenhan.

Lee and I had been chatting for a few minutes when he casually reached up and removed a box stuffed with papers from the top of his filing cabinet. He fingered through files, stopping at a fat folder bursting with pages.

I blinked. Realizing what it contained, I couldn't believe my luck — if I was right, this treasure trove would be almost as good as being able to interview Rosenhan himself. Pages peeked out from a folder titled ON BEING SANE and another marked PSEU-DOPATIENTS. Papers stuck out in various directions. The files were organized, or rather disorganized, according to how Rosenhan left them — once I started pawing through it I quickly realized that the mess revealed more about his mind than anything sanitized by an archivist. There was something voyeuristic, even indecent, about the digging, but, for better or worse, my years working in a tabloid newsroom weaned me off any shame about going through people's dirty laundry.

Sometimes the contents corresponded to the description on the folders; often they did not. You'd open up a folder on, say, Rosenhan's work about altruism in children and you'd find a bill of sale for his Mercedes. There were drafts of "On Being Sane in Insane Places," which Rosenhan had cut out into sections and pasted back together like an elaborate puzzle, and dozens of pages of handwritten diary entries from his time inside the hospital. A folder marked criticism held brutal comments from his peers: "pseudoscience presented as science," "unfounded," "entirely unwarranted." If this folder was any indication, Rosenhan clearly had pissed off psychiatrists. And he seemed proud enough of it to keep the evidence.

I came to a stack of paper held together by a thick but weathered rubber band. The first page read:

Chapter 1

We never really know why ideas are born. Only how and when. And while origins hardly matter when an idea is fully formed and articulated, they may make something of a difference when it is still being shaped. What stands in tonight's shadows sometimes mars tomorrow's path.

I find myself unable to say why this research began in any sense that reveals to me something more about the ideas. Perhaps you, better than I, can infer something more from the circumstances. Let me describe them.

His unpublished book. There were at least two hundred pages here. My heart raced. This was the manuscript that his publisher, Doubleday, had sued him for. These were the pages they fought for but never received — pages the world had never before seen. I tried to look casual as I set it aside and continued my frantic search for information. I wouldn't be able to rest until I understood the study inside and out, including what led to its creation and the context of its consequences. I wanted to be inside the heads of everyone involved. And here was my chance. I tried to contain my enthusiasm when I opened up the folder marked PSEUDOPATIENTS.

My Rosetta stone. The names of all the pseudopatients.

- David Lurie, pseudopatient #1, was a thirty-nine-year-old psychologist who pretended to be an economist and got himself admitted for ten days to Billington State Hospital. He was released with the diagnosis of schizophrenia, schizoaffective type, in remission.
- John and Sara Beasley, pseudopatients #2 and #3, husband and wife, psychiatrist and psychologist, went undercover. John went in twice, first at Carter State for three weeks and then at Mountain View for two. John described his time inside as "Kafkaesque." Sara admitted herself to Westerly County and spent eighteen days inside. Both were released with a diagnosis of schizophrenia in remission.
- John's sister, Martha Coates, pseu-

101

dopatient #4, was a widow who posed as a housewife. She joined the study after her brother and sister-in-law and spent two weeks at Kenyon State Hospital, where she became the fourth pseudopatient in a row to receive a schizophrenia diagnosis.

- Laura and Bob Martin, pseudopatients #5 and #6, followed. Laura, a famous abstract painter, was admitted to the only private psychiatric hospital in the study. She spent a shocking fifty-two days there until she was released with a different diagnosis than the rest: manic depression. Her husband, a pediatrician, admitted himself to a less-than-stellar psychiatric hospital, claiming to be a medical technician. He, too, was diagnosed with schizophrenia.

- Carl Wendt, pseudopatient #7, went undercover four times, totaling seventy-six days locked away. His obsession with the study worried Rosenhan, who became concerned that Carl had grown "addicted" to it.

- Finally, there was Bill Dixon, #8, Rosenhan's graduate student, who infiltrated a failing public hospital for seven days and also received a diagnosis of schizophrenia, making the total seven out of eight patients to receive that diagnosis. All twelve hospitalizations had resulted in misdiagnosis.

It didn't take long to figure out that pseudopatient #1, David Lurie, was really David Rosenhan, which led me to the swift realization that all the names had been changed. There would be no simple, ten-minute internet search for Bill Dixon or Martha Coates. The hospitals, too, had been renamed.

Lee's voice yanked me back to the present moment in his Stanford office.

"David was in some ways a little hard to know," he said.

"What do you mean?" I asked.

"Well . . ." Lee paused here, choosing his words carefully. "He had secrets, in other words, as most people do. It was the dramatist in him. He was, as that saying goes, a riddle wrapped in a mystery inside an enigma."

In retrospect, I wish I had asked him exactly what he meant. But in the moment, I was too distracted by the promise of the pages in front of me.

Lee pivoted back to the files. "You may find the answers to your questions in this," he said, gesturing to the papers. But then he added: "Where's that one thing?" He searched through the pile, stopped at one folder, removed it, and walked it back to his filing cabinet. "This is personal," he said. He placed the folder in his cabinet, closed the drawer, and smiled at me. Was this smile an invitation? Or was I reading too much into all of this?

It was only when I walked back to my car that Lee's words began to circle in my brain: *riddle, mystery, enigma.*

■ ■ ■ ■

PART TWO

■ ■ ■ ■

Felix Unger: I think I'm crazy.
Oscar Madison: If it makes you feel any
  better, I think so too.
                    — *The Odd Couple,* 1968

Part Two

6
THE ESSENCE OF DAVID

I returned to California six months later to revisit the files, which had been relocated to their intended owner, a clinical psychologist and a close friend of Rosenhan's named Florence Keller. Florence had saved the files in the frenzied aftermath of Rosenhan's disabling strokes when Rosenhan was being moved to an assisted living facility a decade before his death in 2012. During the frantic cleanout, Florence managed to salvage a box marked ON BEING SANE. When Florence alerted Rosenhan, he asked that she hold on to it for him.

Florence is trim and attractive, a handsome woman in her early seventies. There is something of Katharine Hepburn in the way she navigates the world — floating with an easy confidence as she swings open the door, welcoming me in with a wide smile. She gave me a tour of her Joseph Eichler–designed Palo Alto midcentury bungalow with its orange and Meyer lemon trees. I noticed two identical *New Yorker* magazines side by side on the kitchen table.

"Why the two?" I asked.

"It's the one thing LaDoris and I can't share," she said, laughing. LaDoris, her partner for over thirty years, goes by many names — "LD" or "Herself," as Florence calls her, and "Judge Cordell" to the rest of the world. She's a Palo Alto celebrity, the first African American female judge to sit on the bench of the Superior Court, who in her retirement now provides legal commentary for the national news and leads protests on all manner of issues, from upholding judicial independence to combating police brutality. If you live in Palo Alto, it's likely that LaDoris either helped you, married you, or advocated for you.

From the moment I removed my shoes and stepped through her front door, Florence and I became partners in crime. I called her my Rosenhan whisperer. She was one I would rely on at every stage of the investigation, through every increasingly surprising twist. She was the person who had the most insight into Rosenhan's mind, and his secrets. The two had met at a mutual friend's party, where she found herself in lively conversation about how almost all curse words aimed at men were really directed at women. The bald man with a glimmer in his eye readily agreed, and the two started listing words that fit her theory.

"Son of a bitch, bastard . . ."

"Motherfucker . . . ," he added.

They each rattled off as many epithets as they could, and by the time they'd run out of insults, the two were fast friends.

I asked Florence to help me translate the dozens of pages of Rosenhan's handwritten notes scrawled out on yellow legal pad paper, written before and during his hospitalization for the study. His handwriting at first had seemed easy and accessible — he had beautiful penmanship — but, strangely, the minute you began to read, you realized that the letters themselves were impossible to decipher. "*Echt* David," meaning "the essence of David," quipped Florence.

Over the coming months, I burrowed into that unpublished manuscript. The study began, I would quickly learn, not with Rosenhan's plan to challenge psychiatry as he knew it, nor even with a Nellie Bly–inspired curiosity about the conditions inside the asylums, but with a student request in his abnormal psychology honors class at Swarthmore College in 1969.

"It all started out as a dare," Rosenhan told a local newspaper. "I was teaching psychology at Swarthmore College and my students were saying that the course was too conceptual and abstract. So I said, 'Okay, if you really want to know what mental patients are like, become mental patients.' "

January 1969
Swarthmore, Pennsylvania

The campus — the whole world, really — seemed to be losing its mind. In the first six months of 1969, there were more than eighty-four incidences of bombings, bomb threats, and arson reported on college campuses. America was mere months away from the national shock of the Manson Family murder spree. Plane hijackings were common. The world had just watched police officers use billy clubs and tear gas on crowds of unarmed protesters at Chicago's Democratic National Convention as onlookers chanted, "The whole world is watching." Richard Nixon's inauguration fell the same week as the start of Swarthmore's spring semester. Some of Rosenhan's students had joined the tens of thousands in Washington who cheered and booed, throwing bottles at the presidential motorcade and holding up signs announcing, NIXON'S THE ONE . . . THE NUMBER ONE WAR CRIMINAL. Nixon, in a moment of inspiration, stuck his head out the top of his limo and made the now infamous V-for-victory sign with his arms. We now know that Nixon's self-serving political meddling helped prolong the Vietnam War, a personal victory achieved by any means necessary. The nightly news showcased the Vietnam War in real time as casualties hit their peak in 1968. We were in an unwinnable war with an enemy on the other side of the earth killing thousands of young men, for *what*? In the face of such inexplicable acts on a global scale, madness no longer seemed to be restricted to the asylums. Some young men who had low draft numbers exploited the system by pretending to be out of their minds to get out of the war. Why not, after all? *Everything* seemed insane.

"It's easy to forget how intense the '60s were," wrote Swarthmore alum Mark Vonnegut (the son of *that* famous writer) in his memoir *The Eden Express,* which chronicled his own experience with psychosis during this turbulent time.

In 1969, the concept of mental illness — of madness, of craziness, of deviance — had become a topic of conversation like never before in the history of our country. It became more of a philosophical debate than a medical one. Wasn't "mental illness," many argued, just a way of singling out difference? Madness was no longer shameful; it was for the poets, the artists, the thinkers of the world. It was a more enlightened way to live. The young embraced psychoanalyst Fritz Perls's slogan (popularized by Timothy Leary): "Lose your mind and come to your senses." Only squares were sane.

And then there were the drugs. Two million Americans had dropped acid by 1970, getting a glimpse of the "other side" and joining the "revolution by consciousness" — convinced, as Joan Didion wrote, "that truth lies on the far side of madness." They did not want what society (their schools, their parents, President Nixon) needed from them. They believed that they were all a razor wire away from the madhouse — and they may well have been.

Young people moved to utopian communities in the middle of nowhere. One of the country's most popular bumper stickers was QUESTION AUTHORITY. *Growing Up Absurd,* written by an openly bisexual anarchist who linked the disillusionment of youth with the rise of corporate America, was a runaway bestseller. The 1966 surrealist film *King of Hearts* featured a small French town during World War I where the happy denizens of the local asylum take over, prompting the viewer to ask, Who is truly sane in a war-ravaged world gone mad? Ken Kesey's trippy novel *One Flew Over the Cuckoo's Nest* did more than any other book to incite the public against psychiatry. (In a few short years, the 1975 movie starring Jack Nicholson would further outrage viewers.) The power of Kesey's story has endured. I'm sure that if someone asked you for an example of a "sane" person railroaded by a mental institution, you would immediately cite *Cuckoo's Nest* as the classic example. Though the book was intended to critique conformity on a grand scale, the novel will forever be associated with the evils of psychiatry. The book, as one psychiatrist put it, "gave life to a basic distrust of the way in which psychiatry was being used for society's purposes, rather than the purposes of the people who had mental illness."

Kesey, a star athlete and the son of a dairy farmer, found his revelatory moment while working nights as an aide at Menlo Park Veterans Hospital. He enrolled in a government-sponsored experiment at the same hospital, where researchers dosed him

with a series of drugs — including mescaline, Ditran, IT-290, and his favorite, lysergic acid diethylamide (LSD).

These experiences birthed the ultimate
antihero, Randle Patrick McMurphy, who
fakes his way onto a ward to get out of serving a prison sentence. "If it gets me outta
those damn pea fields I'll be whatever their
little heart desires, be it psychopath or mad
dog or werewolf," McMurphy says.

Once free of his prison sentence, McMurphy causes as much trouble as he can on
the ward and in doing so discovers that his
fellow patients aren't so different from him
after all: "Hell, I been surprised how sane
you guys all are," McMurphy tells the other
patients. "As near as I can tell you're not
any crazier than the average asshole on the
street." The big difference, McMurphy is
shocked to find, is that the other men
shackled themselves to the institution voluntarily. They *chose* to be there.

Harding, one of the patients, explains
why: "I discovered at an early age that I was
— shall we be kind and say different? . . . I
indulged in certain practices that our society
regards as shameful. And I got sick. It
wasn't the practices, I don't think, it was
the feeling that the great, deadly, pointing
forefinger of society was pointing at me —

and the great voice of millions chanting, 'Shame. Shame. Shame.' " He wasn't sick in the biological sense, but was made sick by the world around him.

Even more pointedly, the narrator, Chief "Broom" Bromden, pretends he can't hear or speak, but documents everything and gets away with it because the institution sees him only as a crazy man with a broom, and so he is invisible. In the end, McMurphy's battle is lost. The authoritative powers of the institution — embodied by the monstrous Nurse Ratched — converge on McMurphy, who is lobotomized for the sake of convenience, never again to be a problem on Ratched's ward.

Suffice it to say, in the early 1970s, psychiatric hospitals were not getting a good rap.

On top of it all, Cold War paranoia touched everyone, as stories of men and women interned in Soviet psychiatric hospitals for political reasons reached the US. Thousands of dissenters in the USSR were hospitalized against their will, including one outspoken general named Pyotr Grigorenko, who served in the Red Army before he began to question the policies of the Communist Party. He was diagnosed with "paranoid development of the personality with reformist ideas rising in the personality, with psychopathic feature of character, and the presence of symptoms of arteriosclerosis of the brain" (a Russian nesting doll of a sentence if I ever heard one). He spent five years in one of the worst Soviet "psycho-prisons" until he was finally released and allowed to immigrate to the US.

Which was scarier: using psychiatric labels as a tool of oppression, or the possibility that many of these Soviet psychiatrists actually believed that someone who didn't support Communism must be crazy?

And yet this exploitation of psychiatry was also happening in America — by the White House, in particular. To discredit Daniel Ellsberg, the man who leaked the Pentagon Papers to the *New York Times,* former CIA agent Howard Hunt sent the "plumbers" (men who did the White House's dirty work) to his psychoanalyst's office to find information there to discredit him.

The most famous person singled out for his mental health history was Republican presidential candidate Barry Goldwater, whom psychiatrists (without personally examining him) called unfit to serve, describing him as, among other things, "a dangerous lunatic" in a 1964 *Fact* magazine article titled "1,189 Psychiatrists Say Goldwater Is Psychologically Unfit to Be President!" The American Psychiatric Association, embarrassed by the resulting fallout (and Goldwater's successful libel suit against *Fact*), implemented the Goldwater rule in 1973, an ethical principle banning psychiatrists from making armchair diagnoses of public figures they have not examined, which continues even in the face of opposition today.[1] A cardiologist, they argue, wouldn't dare diagnose someone they saw only on TV, so neither should psychiatrists. This rule suggests that psychiatry should be held to the same standards as other medical specialties, a defensiveness that is revealing: "Psychiatrists are medical doctors; evaluating mental illness is no less thorough than diagnosing diabetes or heart disease," the APA wrote.

At the same time, the lay public continued to wonder, *Does madness even exist?* This might seem like an absurd question to anyone who has lived with mental illness — either personally or through a loved one —

1. The APA reiterated its dedication to the Goldwater rule in 2018 in response to public debates over President Donald Trump's mental fitness, writing, "A proper psychiatric evaluation requires more than a review of television appearances, tweets, and public comments."

but in a time when people were labeled "mentally ill" simply for their attraction to people of the same sex, it was a legitimate debate. The emerging anti-authority movement questioned so many of our assumptions, arguing that all madness was a social construct. They quoted French philosopher-historian Michel Foucault's *Madness and Civilization* as proof that psychiatric institutions had, from the very beginning, used confinement as a tool for domination. Sociology professors taught the labeling theory, which presented mental illnesses as self-fulfilling prophecies hoisted upon us by society's own need to classify and stereotype "deviants."

If this sounds familiar, it is because these are the same impossible questions (in different contexts) that we've been circling as long as we could reason. And Rosenhan would crystallize all of this in his blockbuster study.

Meanwhile, the growing anti-psychiatry movement launched critical attacks from within the academy's own ranks. R. D. Laing, a Scottish psychiatrist, offered arguments that were most appealing to the counterculture. He theorized that insanity was a sane response to an insane world. Schizophrenia, Laing would write, was a super sanity — a kind of insight only those with truly open minds could achieve — and he believed that one day, "They will see that what we call schizophrenia was one of the forms in which, often through quite ordinary people, the light began to break through the cracks in our all-too-closed minds."

In 1967, he wrote, "Madness need not be all breakdown. It may be breakthrough." Students carried dog-eared copies of his books *The Divided Self* (1960) and *The Politics of Experience* (1967) — two of his most popular and groundbreaking works — in their back pockets, a badge of honor advertising their cynicism about the societal judgments imposed on the mind, proclaiming their higher consciousness about the self, about sanity, about society. But it was easy to poke fun at him. "Schizophrenics were the true poets," Erica Jong would joke in *Fear of Flying.* "Every raving lunatic was Rilke." Soon enough, reports of rampant drug use at Laing's asylum-style London house called Kingsley Hall emerged. Alongside his rise as a guru, Laing seemed to grow into a caricature of kookiness as he flirted with "rebirthing" sessions and other bogus '70s-era treatments, along with copious drugs and alcohol. (I'll never be able to purge the sight of Laing, red-faced and perspiring as he mimed pushing himself through "his mother's birth canal" on a patterned couch, captured on video and screened for me by his former cameraman.)

Hungarian-American psychiatrist Thomas Szasz called mental illness a "myth" and said that the concept of mental illness was "scientifically worthless and socially harmful." The opening of his most famous book, *The Myth of Mental Illness,* reads, "There is no such thing as mental illness," and the book relegates psychiatry to the realm of alchemy and astrology. Psychiatrists used medical jargon, he argued, without having any real credibility. "If you talk to God, you are praying; if God talks to you, you have schizophrenia. If the dead talk to you, you are a spiritualist; if you talk to the dead, you are a schizophrenic," he wrote. Institutional psychiatry in particular was an instrument of oppression to control troublesome or morally deviant characters, whom he called "parasites." Psychiatry wasn't just oppressive, it also enabled the worst among us, he argued. At least for a time, Szasz's arguments were compelling to intellectuals in and outside the field. (According to Rosenhan's private notes, he was far more inspired by Szasz's view of mental illness than

Laing's — at least at first. In later retellings, however, as Szasz fell out of favor he would credit Laing with inspiring his famous study.)

The anti-psychiatry movement made not-so-strange bedfellows with the civil rights movement. Both were united against a common enemy: the power of "the institution" that decided what was "normal" or "acceptable" in society.

This spirit fully permeated Rosenhan's Swarthmore College, an ivory tower liberal enclave with Quaker roots, surrounded by blue-collar, conservative, meat-and-potatoes Delaware County, Pennsylvania. In the spring semester of 1969, the campus had never been so politicized. Though typical university controversies still existed — like whether the admissions office should maintain its ban on students with beards working as tour guides — now they were conducted alongside contentious debates over whether or not to allow naval recruiters on campus.

In the midst of these protests, the Swarthmore Afro-American Student Society (SASS) staged sit-ins and walkouts calling for greater representation of black students on the campus that had opened its doors to them only two decades earlier, and whose minuscule numbers had barely hit double digits. With tactics that included hunger strikes, the SASS successfully delayed the opening of Swarthmore's spring semester, resulting in a week of canceled classes dubbed "The Crisis of 1969," which ended only when President Courtney Smith suffered a fatal heart attack in a campus stairwell. One writer suggested that President Smith died "from a broken heart." The campus mourned the popular president's death, and the Afro-American Student Society's terms were back-burnered. Swarthmore became known as "the place where the students killed the president"; Vice President Spiro Agnew is said to have nicknamed it "the Kremlin on the Crum" (the Crum are the woods that surround the college). Needless to say, the atmosphere on campus that spring was electric.

And these trade winds helped steer a delegation from David Rosenhan's abnormal psych seminar to approach him in his smoky lab in the basement of Swarthmore's Martin Hall at the start of the spring semester in 1969 — a meeting that would set in motion a chain of events that would change the world.

7
"Go Slowly, and Perhaps Not at All"

Professor David Rosenhan may have only just arrived the previous semester, but in his tweed jacket with leather elbow pads, he arrived at Swarthmore looking like he belonged there. Some students joked that his big bald dome of a head must mean *he had a big brain.* Colleagues recalled his hip-swishing swagger as he'd amble through the campus with his hands clasped behind his back, the walk of a guy who owned the place.

Rosenhan's previous position had been as a lecturer at Princeton's Department of Psychology and a research psychologist at Educational Testing Service, a group of test makers that helped shape the SAT into the test we know today. The Educational Testing Service gave its researchers a wide latitude to explore nearly any subject. It was a perfect situation for Rosenhan, who had an agile mind that tended toward backflips, always ready to vault over and around obstacles in his path. (He nimbly employed psychological tricks even in grade school. Rosenhan was a scrawny kid who loved wrestling, and he figured out a way to use his weakness as an advantage. To break an opponent, he'd set the other boy's expectations low by purposefully tripping on the way to the mat.)

The elastic nature of his mind reveals itself in the subjects he pursued: He wrote papers on dream analysis, on hypnosis, and on contemporary social issues like the motivations of Freedom Riders, black and white civil rights advocates who traveled on buses together in the South to challenge segregation. He replicated Stanley Milgram's 1963 study on obedience, showing the extreme lengths that his subjects would go to when following orders. Milgram had created a fake shock box with levers marked with voltages that ranged from 15 V to "XXX," the latter's abstraction meant to imply that it was so high it could be deadly. Milgram's results stunned the world: The study's volunteers showed themselves ready and willing to administer high levels of electric shocks to strangers, just because they were asked to (in Milgram's sample, 70 percent would shock their cohorts at the XXX level), which struck an uncomfortable chord in the aftermath of World War II. The son of two Eastern European Jews, Milgram had grown up in the shadow of the Holocaust, as had Rosenhan, and this was never far from their minds. "A number of us here are interested in extending your work," Rosenhan wrote to Milgram in 1963. "Needless to say, we feel you've discovered a remarkable phenomenon."

Rosenhan's current passion — and the interest funded by the National Institute of Mental Health — was studying pro-social behaviors in children, specifically testing "young children's unprompted concern for others," which he called his "search for values." In other words, do you become a good or bad person, or are you born that way? This was an animating question for social psychologists at the time — one that Milgram and his shock machine and later Zimbardo and his prison experiment both grappled with.

Rosenhan set up his lab to resemble a miniature bowling alley, with marbles used as bowling balls. He rigged the study so that he could control whether a child would win or lose, and then documented how the child's altruistic behavior, like donating money to charity, changed depending on whether or not adults were present. Rosenhan's research assistant Bea Patterson remembers cringing at his instructions to tell the children that they were "duds" if they didn't win, knowing full well that the results were randomly assigned. Sometimes the losing kids would cry. More often they would cheat, pushing over the tiny pins. In an unanticipated turn of events, Rosenhan and Patterson discovered that cheating, as much as winning, increased the likelihood that children would donate their money. Other researchers may have thrown in the towel, but Rosenhan, as any good scientist would, turned his study on its head and published another, more interesting paper about the role of confidence in cheating behavior, an example of his backflipping brain at work.

His intellectual range was boundless. He devoted a good deal of interest to abnormal psychology, and he wrote two textbooks on the subject with close friend and psychologist Perry London. He explained his attraction to the topic in a letter to a colleague and friend: "Abnormal psychology is a painfully complicated psychological area. It implicates biology, chemistry and genetics heavily. It implicates social perception. And it implicates the experience of any of us who has been depressed, anxious, or worse. The need to bring simplicity and understanding to an apparently complicated area challenges me."

But Rosenhan's real talent was teaching. Rosenhan had a way with people, a seductive quality. His baritone voice could easily transfix a packed auditorium. Ex-students of his called it a gift. One described him as being able to "rivet a group of two to three hundred students with dynamic lectures that are full of feeling and poetry and personal anecdotes."

It was no wonder, then, that Rosenhan's first abnormal psychology class was such a hit that Swarthmore tapped him to run an honors seminar devoted to the same subject. I wish I could have been there in the moment to hear him on that very first day, but instead, I was able to track down a few tapes of Rosenhan's later lectures. His deep and resonant golden voice, which sounds a bit like Orson Welles's, boomed through my computer speakers: "We are here in this spring quarter to see if we can understand the mind through its abnormalities," he said. His Talmudic cadence — the way he elongated words, pausing and stressing them for dramatic effect — must have been carved in him during a youth spent singing and training to be a cantor. It was the kind of voice that projected authority and made you want to lean in, focus, and listen.

"The question is . . . What is abnormality? . . . What are we here for?" he asked. "Some things will be black . . . Others will be white. But be prepared for shades of gray."

I had no idea how shaded that gray would become.

It was likely late morning when his students approached him in his office. They had come to complain, he explained in his unpublished manuscript, "that the course had had two shortcomings. First, I had avoided case histories of psychiatric patients. And second, I had pointedly refused to allow students to visit psychiatric hospitals." He went on:

We sometimes forget that psychiatric patients are people too. They have their dignities, their shames, and their vulnerability like the rest of us. It seemed unfair, an invasion of the privacy of people who were helpless to defend themselves, to encourage students to visit such hospitals. Would you want to be exposed to young inquisitive strangers, however well intentioned, if you were there? . . .

For their part however, the students had a case and they pressed it vigorously. We do not appreciate abstractions, they argued, without direct experience with the substances that form them. How does one assess . . . say, schizophrenia, without knowing directly some schizophrenics? Without having been exposed immediately and concretely, to their thoughts, their feelings, the way they perceive the world? Isn't it a bit like trying to understand the value of a dollar without knowing what the dollar will buy?

I was caught then, clearly and unpleasantly, between appreciating their views and being convinced by my own. As the issues became clearer, the argument took on vigor. Finally, it seemed to me that I saw a compromise between these two seemingly irreconcilable positions.

"Look," I blurted, "if you really want to know what psychiatric patients are all about, don't waste your time on case histories or in simply visiting hospitals. Why don't you simply check into a psychiatric hospital as a patient?"

"When?" they asked.

When? Not why. Not how, or where, or

even "hey wait a minute." But when. Bless their cockiness.

As his students made their case, Rosenhan recalled an undergraduate course at Yeshiva University on minority groups, which required each student to rent a bed in a Spanish Harlem boardinghouse to experience poverty firsthand. Living with ten others in an apartment meant for four people had made a deep impression on Rosenhan, even as the son of Polish Jewish immigrants in Jersey City who survived on his father's meager living as a door-to-door salesman. The memory rekindled an enthusiasm he recognized from his own student days.

Energized, he decided to reframe the students' pitch as a teaching exercise and began to plan. First they'd have to find a psychiatric hospital willing to let them in. Luckily a colleague worked at Haverford State Hospital just fifteen minutes away, and he promised to bring it up with the hospital's superintendent, Jack Kremens. Rosenhan couldn't believe his luck. Kremens, who had worked during World War II as an agent in the Office of Strategic Services (a precursor to the CIA), would be the perfect point person to approach for something so bold.

And he had every reason to think Kremens would be interested, too, since the students' undercover exercise would allow them to report back, from the ground, about internal operations at the hospital. Rosenhan and his students could document any gap between the set of regulations regarding patient care and the day-to-day realities. Kremens had been specifically concerned about the possibility of illegal drugs floating around his facility, and he needed to know if they were coming from someone on the inside. Rosenhan's project offered an opportunity to do some spying.

But there were some serious downsides to going undercover at Haverford, too, given exactly those conditions. Three years later, in 1972, a Haverford Hospital nurse named Linda Rafferty would sue the hospital, exposing a host of offenses, including "homosexual abuse by other patients; . . . sexual exploitation by outside workmen; . . . leaving blank prescription forms, signed in advance by physicians, in unlocked drawers for nurses to fill out on weekends; and chronic absenteeism on the part of the hospital's medical staff."

Though Rafferty's allegations were on the extreme end, it was a precarious time for all psychiatric hospitals, as they were in the

131

midst of profound changes — none more transformative than the new drugs now flowing through patients' bloodstreams. Chlorpromazine (marketed under the name Thorazine in America) seemed at the time to be psychiatry's pivotal twentieth-century discovery. It hit the American market in 1954 and by the end of the next decade had infiltrated most psychiatric hospitals. Thorazine was, as historian Edward Shorter put it, "the first drug that worked" and, according to psychiatrist, psychopharmacologist, and vocal critic of the pharmaceutical industry David Healy, "widely cited as rivaling penicillin as a key breakthrough in modern medicine."

Chlorpromazine came from a happy accident: After a researcher tested the antihistamine out on rats and found that they were uninterested in climbing a rope to get their food, French naval surgeon Henri Laborit tested the drug on surgical patients and found that it had a dissociative, sedating effect. *Open me up, who cares,* seemed to be the vibe. Why not, his peers wondered, try this drug out on psychotic patients?

The results were astounding, though not uncontroversial. In a remarkable number of patients, the most pronounced positive symptoms of schizophrenia — the hallucinations, paranoia, and aggression — faded away. Journalist Susan Sheehan describes the miracle of Thorazine in her 1982 book, *Is There No Place on Earth for Me?* "Thousands of patients who had been assaultive became docile. Many who had spent their days screaming subsided into talking to themselves. The décor of the wards could be improved: chairs replaced wooden benches, curtains were hung on windows. Razors and matches, once properly regarded as lethal, were given to patients who now were capable of shaving themselves and lighting their own cigarettes without injuring themselves or others or burning the hospital down." Pharmaceutical companies added other related drugs with brand names like Compazine, Stelazine, and Haldol by 1969, the year Rosenhan went undercover. A year later, antipsychotics were minting money for the American pharmaceutical industry to the tune of $116.5 million (which today would be $780 million) a year.

This started the modern, drug-dependent era of psychiatry. Psychiatrists might not have been able to find and identify the "seat of madness," but now at least there was a way to treat it, wherever it was. Other breakthroughs soon fell in line: the discovery of antidepressants, lithium for bipolar disorder, and Miltown for anxiety. Though little was yet known about brain chemistry (depression was still viewed by many as "inward-directed anger," obsessive-compulsive disorder as "arrested psychosexual development in an anal stage," and schizophrenia as the result of overbearing mothers), psychiatry now had an armamentarium and a language — *take that, oncology!* — that gave it legitimacy as a true medical specialty. Later, as more insight into brain chemistry emerged, our terminology changed. We developed schizophrenia because of a "dopamine disorder." We were depressed because of a "catecholamine disorder" (later a "serotonin imbalance") and anxious because of a "5HT disorder." It all appeared so comfortingly scientific, and the public embraced this new insight into our minds/brains. And with this insight came new ramifications for misdiagnosis: Different drugs treated different conditions (antipsychotics, like Thorazine, were diagnosed for people with schizophrenia; mood stabilizers, like lithium, for manic depression; and antidepressants for those with depression). Diagnostic mistakes suddenly *meant something.* There was now a premium on diagnosis — not only for doctors and patients, but for insurance and pharmaceutical companies, too.

Despite the obvious progress, it wasn't a smooth transition, however. Kesey documented the array of drugs — and the backlash to them — in *One Flew Over the Cuckoo's Nest:* "Miss Ratched shall line us against the wall, where we'll face the terrible maw of a muzzle-loading shotgun which she has loaded with Miltowns! Thorazines! Libriums! Stelazines! And with a wave of her sword, *blooie!* Tranquilize us all completely out of existence." Though the general effectiveness of the medications was unmistakable — even if they were perhaps *too* effective, as shown by Kesey's quotation — many psychiatrists insisted that they offered a skin-deep fix that did not address the all-debilitating deficits that had diffuse effects across a wide range of ordinary life situations.

Once Jack Kremens had agreed to host undercover undergraduates at Haverford, despite all the risks, Rosenhan and his students discussed the specifics of the study. Would the staff be aware of their presence or not? Would they make up names or use their own? What addresses would they use? Most crucially: How would they get out

135

once they got in?

The first few decisions came easily. They would change their last names and keep their given names. The students would identify as such, but claim to be from different universities to protect their anonymity. (After all, how many potential employers would believe you if you said: *Oh yes, I was institutionalized, but it was for a class . . . ?*)

It may have started as a dare, but it quickly morphed into something more provocative — a teaching exercise. Though the superintendent knew about their mission, Rosenhan made certain that the rest of the staff remained in the dark. So they still needed to convince the hospital that they required help. What symptoms would get them in? This became the source of debate. Would the pseudopatients chew up the scenery pretending to be mad — wide eyes, dirty clothes, ranting and raving the way Nellie Bly had — or would they play it cool? What did madness look like anyway?

"We were all keyed up," Swarthmore student Harvey Shipley Miller recalled. "I certainly was. I'd never been inside [an institution]. This was exciting."

They came up with auditory hallucinations — *hollow, empty,* and *thud* — words

136

that practically screamed of ennui, an existential crisis. Frankly, this should have raised an immediate red flag at the institution because, according to Rosenhan, there had been exactly zero cases of existential psychosis reported in the literature. Rosenhan joked in a letter to a friend, "They will probably write a paper about it!" In a very obvious way this choice thumbed a nose at the rube psychiatrist who most likely had never read much Kierkegaard — the Swarthmore version of an inside joke. At this point, according to his manuscript, Rosenhan had no plans to publish anything himself or collect serious data. Their one goal was to get into the hospital by any means necessary with as little risk to the students as possible.

They studied the work of the few academics who had attempted similar coups before them, among them medical anthropologist William Caudill, who lived for two months in 1950 as a patient of a psychiatric hospital associated with Yale, writing up his traumatic experiences in the article "Social Structure and Interaction Processes at a Psychiatric Ward." Caudill exaggerated his own issues at his intake interview, amplifying his marital troubles and intensifying his anger and alcohol issues, but he kept the rest of his biography intact. Still, Caudill claimed that even such minimal lying took a serious toll on him, generating deep inner turmoil about having to live as an impostor. It got so intense that Caudill warned against any replications. One of his supervisors who visited him in the hospital commented, "I believe he lost his objectivity as a participant observer, and almost became a participant, a patient." Rosenhan made a note of this in his own writings and, unlike Caudill, vowed that the participants would "not alter our life histories in any way, nor describe pathology where none existed in our current lives, or exaggerate our real problems."

Rosenhan and his class read exposés by journalists from around the country who had, like Bly before them, revealed the barbarity occurring in our backyards. During World War II, three thousand conscientious objectors were assigned alternative service at state psychiatric hospitals around the country. Shocking photographs taken by one of the objectors were featured in Albert Maisel's "Bedlam 1946," published in *Life* magazine. Maisel's article described brutal conditions inside Pennsylvania's Philadelphia State Hospital at Byberry and Ohio's Cleveland State Hospital — beatings so bad that people died — alongside those deeply disturbing photographs that looked uncomfortably close to images that had just emerged from liberated German death camps. In one, a patient sits on a wooden bench, arms mummified by a white strait-jacket revealing legs riddled with untreated sores. In another, a group of men huddle, heads down, naked on a refuse-covered floor.

This was a deranged version of *Groundhog Day* — the same atrocities repeated time and time again. Harold Orlansky compared American asylums to Nazi death camps in his "An American Death Camp," published in 1948. Frederick Wiseman's damning documentary *Titicut Follies* documented in stark black-and-white the forensic (for "the criminally insane") hospital Bridgewater, where the patients were physically and verbally abused — all in front of a camera. Men wandered the hospital grounds naked; a man in solitary confinement banged his head and fists against the wall, spraying dark black spots of blood. An Eastern European psychiatrist interviewed a pedophile, asking questions like: "What are you interested in, big breasts or small breasts?" In one of the more unwatchable scenes, the same psychiatrist smokes while force-feeding a man using a rubber tube, the ashy end of his

139

cigarette perilously close to the funnel. These were dramatic, appalling stories, but they lacked a key ingredient necessary for wide-scale change: They weren't *scientific.* Ultimately, it would be Rosenhan's own study that would slide in and fill that void — though he and his students had no idea of the power of this idea at the time.

Rosenhan was most inspired by the work of sociologist Erving Goffman, who spent a year undercover as an assistant to the physical education instructor at St. Elizabeths Hospital in Washington, DC, all the while recording the inner workings of the deeply dysfunctional mini-city of six thousand patients. In *Asylums,* his famous text published in 1961 (a big year for landing punches, the same year Laing's second book, *Self and Others,* and Szasz's *The Myth of Mental Illness* hit the shelves), Goffman described the hospital as a "total institution," much like prisons and concentration camps, that dehumanized and infantilized patients (really prisoners) and not only did not effectively treat but actually *caused* the symptoms of mental illness. Institutional life not only didn't cure mental illness but actually contributed to chronicity, a condition that psychiatrist Russell Barton named "institutional neurosis" in 1959. Though

Asylums was a groundbreaking work and remains highly respected within sociological and psychological circles, it did not reach the masses in the same way that Rosenhan's paper would.

To his students, Rosenhan assigned work that described psychiatric hospitals as "authoritarian," "degrading," and "illness-maintaining," among other terms. Clearly, he did not expect to find a great deal of healing going on inside those walls.

Perhaps this was why Rosenhan required that the students receive permission from their parents to participate in the study, even though the students were over eighteen years old. Parent responses were far from supportive. "Wasn't it dangerous?" they asked. "How could one be sure that real patients would not harm the pseudo-patients? What about staff? It had been said that occasionally staff are hurtful and worse to patients." How would Rosenhan ensure that the pseudopatients would not be "molested or harmed" from "shock therapy, even lobotomies, not to speak of medications that might be poured or injected into them?" One mother flatly refused, explaining that she had been an employee at a psychiatric hospital and she would never trust her son in the care of one. Another summed it up with one sarcastic sentence: "I hereby give you permission for my son to participate in your ~~insane~~ experiment on insanity."

Rosenhan noted that the parents had all reached the same consensus: "Perhaps hospitals cure, but psychiatric hospitals don't. They brutalize, torture: they are outside the pale; they make the sick sicker, and even the sturdiest, sick."

They make the sick sicker.

Rosenhan contacted a friend, psychiatrist Martin Orne,[1] for advice, who responded: "Go slowly, and perhaps not at all."

History made it clear. Psychiatric hospitals were far from therapeutic. David Rosenhan couldn't subject his students to being committed to one of those hospitals without first seeing what they were up against.

First, he would have to go in alone.

1. Dr. Orne would later make waves himself when he released the transcriptions of his therapy sessions, conducted between 1956 and 1964, with poet Anne Sexton to her biographer seventeen years after her suicide.

8
"I Might Not Be Unmasked"

Rosenhan pulled from his real-life experience to make a kind of bizarro-David, one with a new last name, address, and occupation. He took on his mother's maiden name and became David Lurie, an out-of-work economist / advertising executive. This would be easy to fake since he had, in real life, pursued a master's degree in mathematics. (He dropped this focus when he didn't rank first in his class. Rosenhan didn't do anything that he didn't feel he was the best at, his son, Jack, has explained to me, so he decided to switch to psychology.) Beyond growing a beard ("lest I be recognized!"), he didn't alter his physical image much, planning simply to wear shabbier items from his own wardrobe.

He went ahead and arranged his visit at Haverford through Kremens, making sure no others on staff would be aware of his ruse. Yet, despite all his bravado, as zero hour approached he began to get cold feet. "Thinking and discussing are not like doing," he wrote in his unpublished book. "I was frankly panicked. Would I actually get in? On the basis of such a simple symptom? I began to have serious doubts not only about my ability to get in, but even about my desire to be hospitalized."

His wife, Mollie, did little to alleviate her husband's worries. And she was not one to keep quiet when things bothered her. They had met on the first day of Rosh Hashanah outside a synagogue in Lakewood, New Jersey, in 1958. The two young lovers got so lost in conversation that they didn't even make it inside for services. When Mollie left Rosenhan's side later that summer to return to the University of Chicago, they exchanged desperate letters. One by Rosenhan read: "Remember how I touched your arm and you touched it and wanted to be touched, so I touched your breast and [you] put your arms around me. I'm thinking I loved you without thinking you loved me back . . . I wanted to receive so greedily and tearfully. It hurts. My, it hurts terribly." Two weeks after their first meeting, Rosenhan boarded a flight to Chicago and proposed. As independent as she was, Mollie desperately wanted a family, having been an only

144

child raised in a crowded hotel. (Both of her parents were innkeepers who catered to wealthy Jews on summer vacation.) She and Rosenhan married and a few years later adopted two children — first Nina and then Jack.

Mollie was the prickly one, the difficult one, the tough one — she was notoriously persnickety about her food and would haughtily return meals at restaurants, never too shy to make her grievances known. Or at least that's how she appeared. Close friends described her as warm and caring with a delicious sense of humor. She was a feminist when that was still a dirty word, and she was a scholar, receiving her PhD in Russian history, teaching college classes, publishing on a wide array of feminist issues, and later co-founding the Stanford Center for Research on Women while also raising the couple's two young children. One of her closest childhood friends shared with me a picture that seems to sum her up: Mollie as a teen on a trip to Israel, sitting in the bed of a truck, holding a semi-automatic rifle.

Mollie appeared to be the force in the couple, but those who knew them well saw something else. Rosenhan knew how to sway her. Though she hated the thought of her husband going into a psychiatric hospital, it didn't stop her from helping him prep for his role.

On Wednesday, February 5, 1969, Rosenhan set the study in motion by cold-calling Haverford State Hospital to ask for help. The phone logs recorded a man who had difficulty expressing himself "as his speech was retarded, and he was very emotional." The idea of Rosenhan's speech being "retarded," or, in more modern parlance, delayed, is laughable knowing the natural and gifted speaker he was. Perhaps his nerves were getting the better of him; perhaps, out of fear that he would be exposed as a fake, he leaned into his acting role; or perhaps the operator expected to hear the voice of a "crazy person" so that's what she heard. Either way, he needn't have worried: The operator was concerned enough about his symptoms to advise that "David Lurie" consult with his wife about coming to the hospital the following afternoon. It was his first test, and he had passed with ease.

Rosenhan had a hard time sleeping that night. By the morning, his dread had shifted into tingly jitters mixed with sudden clear-headedness of purpose. He put on an old raggedy button-down shirt, worn gray flannel slacks, a moth-eaten beige pullover, and tired Clarks that had long served as his weekend gardening shoes.

If Rosenhan glanced at the *New York Times* that morning during breakfast, he might have noticed this story: Two court-martialed soldiers were held in a sanity inquiry for mutiny after taking part in a sit-in demonstration. A psychiatrist had testified that the soldiers, who allegedly led the mutiny, were sane — but that they both "suffered impairment of their ability to do what was right by society's rules because both [have] sociopathic tendencies." But did this make them crazy? The jury was still out.

If sanity and insanity exist, how shall we know them?

It was time for Rosenhan to commit himself to the mental hospital.

Like all of us, Rosenhan didn't or couldn't share some things even in his private writing. Through his son, Jack, I learned that Rosenhan's younger brother struggled with manic depression (now called bipolar disorder). Rosenhan's family home was a rigidly Orthodox one, and as his younger brother came of age he grew even more conservative — becoming Ultra-Orthodox, the opposite of David, who may have studied the

Torah as a hobby but approached Judaism with a scholar's eye more than as a true believer. His brother's extremism capsized other aspects of his life. He had difficulties with money, for example, and during manic phases when off his medications would often call Rosenhan to discuss his finances, issues with his growing family, and his various paranoid fixations that this or that person was out to get him.

"My dad was constantly on the phone with his brother dealing with that and trying to help with that," Jack said. "I would hear my father being upset and just saying when he's on his lithium he's fine, but when he's not he has these manic episodes and these grandiose ideas. Eventually [because of] one of those ideas he moved his entire family to Israel." Jack believed that these experiences with his brother shaped Rosenhan's interest in psychology — especially abnormal psychology — and contributed to his zeal for reform, but Rosenhan never discussed this family issue publicly.

On the late-winter morning of February 9, 1969, Rosenhan and Mollie climbed into their VW hatchback, leaving five-year-old Jack and seven-year-old Nina, both of whom were blissfully unaware of their father's plans, with a babysitter. A new worry had cropped up, overriding even the fear of exposure: "a fear that I might *not* be unmasked." Rosenhan handled the stick shift as his thoughts raced: "Do I need shirts, ties, and underwear, or will I be wearing pajamas all day? Or will it be government-issued clothes? Do I need a heavy sweater for the cold days? Will I be going out at all? The children were in school. Will I be permitted to call them? Do they even have phones on the ward? Will they allow me to smoke, and could I bring my lighter?"

The Rosenhans drove through the Philadelphia Main Line. Stately mansions with pristine lawns lined the way. A semicircular gray stone wall provided the only indicator that they were entering Haverford State Hospital's manicured grounds. They drove to the five-story redbrick admissions building, aka Building Four.

No wonder people called it the Haverford Hilton. Built just seven years before Rosenhan's visit, in 1962, Haverford Hospital was an outlier in Pennsylvania in that it was *new* — few states were allocating funds to building psychiatric hospitals. A psychiatrist who worked there described a large recreation building with a gym, billiards room, pool, barbershop, beauty salon, and soda fountain. There was a four-hundred-seat auditorium, bowling alley, library, and fully equipped surgical unit with X-ray equipment, an operating room, and a high-speed sterilizer (cutting-edge at the time).

It was "the Queen Ship," a shining example of the next generation of psychiatric hospitals. Back when Haverford State was being built, a project designed to address the overcrowding in nearby Norristown State Hospital, the construction was delayed five years as neighbors protested the placement of a mental hospital (no matter how groundbreaking) so close to their expensive properties. In response, Superintendent Jack Kremens went door-to-door, introducing himself to convince the community that the hospital would not be a danger or an eyesore, but a welcome addition to the community. He not only got approval but even managed to sign up a few neighbors as volunteers. After it was built, Kremens proudly called it his own "showpiece of radical design," the first of its kind in the world, he told reporters.

Kremens was being hyperbolic, however. It was really the second of its kind. Five of Haverford Hospital's buildings, which catered to long-term hospitalizations, were modeled off the revolutionary work of British psychiatrist Humphry Osmond.

150

Osmond, a "guru of the 1960s psychedelic movement" who is credited with bringing LSD to the mainstream of scientific research, was among the first to study similarities between the effects of psychedelics and psychosis. During Osmond's psychiatric residency, he chanced upon a paper written by chemist Albert Hofmann, who had described the effects of the new chemical compound lysergic acid diethylamide (LSD) in 1943 after ingesting trace amounts of it, resulting in a whopper of a bike ride. Osmond recognized Hofmann's symptoms — depersonalization, hallucinations, and paranoia — in the presentations of schizophrenia he'd seen in his residency. He speculated that maybe LSD affected the brain similarly to the way schizophrenia did — a new theory of the neurobiological cause of mental illness during a time when psychoanalysis still dominated the field. Armed with this brain chemical theory, Osmond conducted a series of experiments dosing psychiatric patients (and — why not? — himself) with LSD and mescaline. He also administered the drugs to alcoholics, other addicts, and treatment-resistant psychopaths with successful results.

Osmond's acid trips also piqued his interest in the environment's influence on the

151

experience of madness, leading to the realization that the way buildings are structured can aggravate or temper positive and negative hallucinations. He argued that most hospitals should be torn down. "They're ugly monuments to medical error and public indifference," he told *Maclean's* magazine in 1957. In his redesign, he made the wards circular to promote greater social interaction, while also adding access to solitary spaces that would allow patients the dignity of privacy.

Osmond gave LSD — which he said allowed one to "enter the illness and see with a madman's eyes, hear with his ears, and feel with his skin" — to architect Kiyoshi Izumi, with whom he was working on a design for a Canadian psychiatric hospital. *To see with a madman's eyes* was a precondition, Osmond felt, to work with or build for him, because, as he wrote in his famous 1957 paper "Function as the Basis of Psychiatric Ward Design": "It would be heartless to house legless men in a building which would only be entered by ladders or very steep gradients," in the same way that it would be heartless to erect a depressing or ominous structure for people who had perceptual or emotional issues.

While under the influence of LSD, architect Izumi traveled to traditionally designed hospitals and found serious flaws for anyone dealing with issues of perception. The patterned tiles that covered the walls confused the eyes. The lack of calendars and clocks created a foreboding timelessness. The recessed closets were so dark that they seemed to gape like open mouths. The raised hospital beds were too high for patients to comfortably sit and touch the floor with their feet — something that seemed to be comforting during psychosis. The long corridors were intimidating.

Osmond agreed, calling the old hospitals "illusion-producing machines *par excellence,* and very expensive ones at that. If your perception is a little unstable, you may see your old father peering at you from the walls." Osmond and Izumi built their ideal mental hospital in Canada, a design that Kremens's Haverford copied. Though Haverford didn't use Osmond's cheese wedge design (creating a double-Y-shaped structure with private rooms, shared sitting rooms, and shared bathrooms instead), the hospital incorporated many of Osmond's theories. Pleasant, uplifting colors replaced patterned tile. The beds were lowered closer to the floor. The furniture was supposed to look like it had come from the patients' own homes. Patients now came first — at least in terms of their immediate surroundings. That is, if you were lucky enough to live in one of Osmond's buildings.

Rosenhan wasn't.

When Rosenhan walked into the admitting room, he noticed that the furniture seemed "used here but not loved." State-issued. Drab. "Not a picture nor an object nor a poster softened its state-owned décor. Clearly purchased at the lowest bid for the minimum specifications . . . it was owned by an anonymous State," he wrote. This was a part of the hospital apparently untouched by Osmond's theories. Rosenhan introduced himself to the receptionist in an almost giddy state, high from the alien sensation of using a name that wasn't his own. When she asked for his driver's license he nearly gave himself away but quickly recovered, saying he'd left it at home. The receptionist moved on to the next question on the form without comment.

Case Number: #5213
Patient name: "Lurie, David"
Address: 42 State Road, Media, PA
Next of kin — name, relationship: Mrs. Mollie Lurie (wife)
Age on admission: 39

Birthdate: 11/2/29
Race: W
Sex: M
Religion: Jewish
Marital Status: Married
Occupation: Advertising writer
Employer: Unemployed
Previous hospitalizations: None

And then they waited.

And waited.

This stoked Rosenhan's irritation. He thought about how Mollie would not get home in time to relieve the babysitter and there was no pay phone in sight to call. *What if I had really been a patient?* he thought.

Then, at a quarter to four, nearly two hours after his appointment, the admitting psychiatrist, Dr. Bartlett, called Rosenhan into his office.

9
COMMITTED

Case #5213 sat on Dr. Bartlett's desk as a reminder that he had left a patient waiting for nearly two hours. That was not unusual. Dr. Bartlett had lost the battle over time management at the hospital years ago.

Dr. Bartlett, hardly ever without a cigarette, read the form: This was David Lurie's first hospitalization.

Lurie walked in. Dr. Bartlett took a beat to assess him physically. He would later describe him as a short and balding man with an academic air, an intellectual type, like a cartoon version of a poet or a struggling professor, with his glasses, beard, beaten-up penny loafers, and weathered khakis.

Dr. Bartlett opened with some basic questions: Name? Age? Date? Location? Bartlett noted that the patient responded slowly. He was clearly uncomfortable, nervous even, but he was oriented.

"I've been hearing voices," Lurie said. Bartlett observed that Lurie grimaced and twitched. The aural hallucinations, Lurie said, started four months ago: "It's empty." "Nothing inside." "It's hollow, it makes an empty noise."

The interview continued for half an hour. Lurie spoke of an inability to choose a path in college, even though he was a successful student. "He has tended to get lost in unproductive creative fantasies and possibly used his intellect to rationalize his failures and lack of progress, professionally and socially," Dr. Bartlett wrote. Lurie also talked about job problems. He shared his shame about borrowing money from his wife's mother, which he said was "embarrassing."

Two pages of richly detailed typewritten notes ended with this conclusion: "This man who is unusually intelligent has had a long history of not directing himself very well, or of fulfilling his potential . . . He is very frightened and depressed."

Dr. Bartlett's diagnosis: schizophrenia, schizoaffective type, defined as a "category for patients showing a mixture of schizophrenic symptoms and pronounced elation or depression."

IMPRESSION:
Schizophrenia, schizo-affective type, depressed 295.74

Dr. Bartlett did not need to commit
Rosenhan. There were excellent outpatient
buildings on the grounds that he could have
recommended. But Dr. Bartlett saw "David
Lurie," a very sick man who needed serious
help, and wanted Mollie to commit her
husband to the facility, effectively handing
over many of his civil rights and allowing
the hospital to hold him for as long as thirty
days. If Rosenhan wanted to leave, he would
have to petition the hospital.

Mollie balked. She told the doctor that
she needed to see her husband alone before
signing anything.

The two huddled in a back corner of the
waiting room, whispering. *Should they call
Jack Kremens? What exactly did voluntary
commitment mean? Would David have to
miss some classes if the hospital refused to
release him before his leave was up? How
about the kids, who knew nothing about any
of this — only that their father was going to
take a short trip? How would they react to his
unexplained absence?* According to Rosenhan's diary, Mollie phoned an unnamed
psychologist friend to get her opinion. The

158

psychologist exploded: "You both are crazy. Him for doing it, and you for letting him."

Mollie charged back into Bartlett's office. *There must be another way,* she pressed. But Bartlett insisted: The hospital allowed only commitments, not voluntary admissions. Lurie must be committed. It was standard procedure. There was no other way of getting into the hospital. Dr. Bartlett argued that it was "really for the patient's own good" and that this was "merely a technicality, nothing to get upset about. That's the way we do things here, and it doesn't really matter."

"Like hell it didn't matter!" Rosenhan fumed. He was particularly upset that Superintendent Kremens had not forewarned them about this procedure. Perhaps, if you're not the one going through them, matters like these might seem merely bureaucratic. But when your own rights — your ability to leave, to refuse medications, to eat and sleep when you want — are on the line, it's a different story.

Rosenhan described a visibly shaken Mollie managing to keep it together long enough to sign. She stopped short at one document that gave the hospital permission to administer electric shock therapy, but permission was mandatory for him to be committed.

Dr. Bartlett assured Mollie that "we do not administer any type of insulin or electric shock without consulting the family first." But this did little to ease the threat. She decided she would not sign this document. Rosenhan grabbed her hand. He needed her. She would be able to visit him every day. Rosenhan did not explain how he did it, but eventually, she signed.

And so began Rosenhan's odyssey into lunacy.

10
NINE DAYS INSIDE A MADHOUSE

Day One

Nurses' Note: 2/6: Thirty-nine year old. Adm to 3 South this PM. History done. First psych admission.

First, the nurse confiscated Rosenhan's belongings — a bag with extra clothes, a toothbrush, and his tape recorder. When she saw this last item, she confiscated it because it was "illegal" and would "disturb the other patients." The nurse left him with his pen (luckily) and five dollars, which she explained was the most that a patient could have. She then told him to strip while keeping the door ajar. Even if this was a safety procedure, she showed no respect for his modesty, as if the moment the system deemed him mentally ill he was no longer entitled to basic human decencies. She took his temperature, his pulse, and his blood pressure — all normal — and measured his

161

height and weight without a word. Even though she was doing all these tests on his body, she acted as if he weren't there at all.

The nurse led Rosenhan into an elevator and up two floors. The elevator opened onto a set of locked, heavy doors. She opened the door with one of her many keys — which clacked as she walked, a sounding bell to guard her against being mistaken for one of them — *him.* Rosenhan stared down the shadowy corridor. He had expected the stereotypical noise of Bedlam to greet him, but all he heard was the metallic banging of the nurse's keys, those symbols of freedom. "Opening the locked door of this unit, you felt as if you were entering a dark foreboding cave where danger lurked," one Haverford psychiatrist wrote in a memoir about his time working on men's 3-South, Rosenhan's new home. "I was often in fear of physical harm."

Rosenhan walked past the brightly illuminated, glassed-in nursing station — aka "the cage," locked at all times — where the nurses could observe the dayroom without having to interact with the patients.

He may have noticed the smell — a sickly sweet aroma of coffee, cigarette smoke, ammonia, and incontinence common to most hospital dayrooms. A patient ran up and enveloped him in an aggressive bear hug. Once the nurse helped extricate him from the embrace, she deposited Rosenhan at a table; his presence — fresh blood! — unsettled the ecosystem, sending the room into a frenzy.

"Son of a bitch!"

"Cocksucker!"

"I only hit him with my open hand!"

These are some of the snippets of dialogue Rosenhan managed to write down as he waited. Most patients were diagnosed, like Rosenhan, with schizophrenia. Some, catatonic, sat staring blankly like the men in the hallway; others paced, muttering to themselves, shaking their fists, or crying out. One psychiatric resident, upon seeing the scene at 3-S, asked, "What the hell have I gotten myself into?"

Rosenhan sat frozen for two hours, his hunger and urge to urinate growing as the feeling of vulnerability immobilized him, something he would later refer to as "the freeze." He realized that he was entirely defenseless. His mind ran in circles: *Where to wash up or to shower? What does one do here? How does one spend one's time? Is there a phone? Can I call my wife and children? When will I see the doctor? When will I get my clothes back?*

163

"For all my sanity and experience, for all that I knew better than others what I was getting into, I was dazed into helplessness," he later wrote.

Someone — likely an attendant — handed Rosenhan a plate of cold, gelatinous stew, a cup of warm milk, and an orange. Rosenhan stared at it in disgust, not realizing that an orange was a rare delicacy inside these walls. Anything edible birthed outside the asylum was a prize.

Day Two
Nurses' Note: 2/7/69 Patient offers no special cps [complaints] during the night. Apparently slept well.

A blaring fire alarm sounded at 6:30 AM.
"C'MON, YOU MOTHERFUCKERS, LET'S GO."
These words greeted Rosenhan his first morning.

He'd had a terrible night's sleep. The sounds of the ward kept Rosenhan in a constant state of fight or flight. Sleep finally came late in the morning, but lasted only until he was jostled awake by a vivid dream of being unmasked. Now in the light of day he had the chance to examine his surroundings. He noticed the spokes of the steel

164

beds, the undressed windows, the bare beige walls with metal night tables standing on beige tile floors, the strange bodies in their identical beds.

Again: "C'MON, YOU MOTHERFUCK-ERS, OUT OF BED."

Rosenhan's roommates stirred, lifting their bodies as if in slow motion. Rosenhan averted his eyes to avoid intruding on these strangers' morning rituals, but was too frightened not to track their movements out of the corners of his eyes. He didn't know anything about these men besides the names yelled at them. Why were they here? Had they done something criminal? Were they dangerous? One of his roommates, a man named Drake, who had lost his mind sniff-ing glue, grabbed his toothbrush and walked by Rosenhan's cot, waving a "hi" as he passed. "He knew I had been watching," Rosenhan wrote.

He shuffled into the bathroom line. Men joked and jostled. Rosenhan hung back, overwhelmed by the smell. The toilets had overflowed. Barefoot patients goose-stepped around the mess, complaining to an at-tendant who watched but did nothing. In the chaos, Rosenhan managed to muscle his way to the double-headed sink. "I looked in the mirror at a bearded, puffy-eyed man,"

165

he wrote in his unpublished book. "I looked as I felt: haggard."

In the cafeteria, Rosenhan, uncertain about the rituals surrounding the meals, watched the others, copying their fluid motions: Remove a plastic tray, pick up a napkin, move steadily down the line, pick up a dish, place it on the tray, sidestep to your left, and repeat. Three lunch ladies stood behind the counter. Their job was to stop any patient from getting too greedy with the food.

"Hey, one butter only," one said.

"You can have another cup after you've finished that one," said another.

"Hey you, get away from there!"

"Desserts are no good for you. They'll rot your teeth."

When Rosenhan sat down, he realized that he had forgotten to grab silverware and an orange. He was too intimidated to return to the line — "the freeze" again.

When he was alone in the hallway or in a quiet part of the ward, he felt he had to constantly monitor his surroundings, eyeing every person, swinging around to catch someone sneaking up behind him. "Tom Szasz is wrong," he wrote, referring to the author of *The Myth of Mental Illness.* "They really are different from me." (Despite be-

166

ing associated with Szasz and the anti-psychiatry movement, Rosenhan complained about being lumped in with them, namely because of their belief that mental illness was not real.)

There was nothing to do except wait. Wait for breakfast, wait for lunch, wait for the doctor, wait for the nurse. If he wanted to smoke — and he did almost constantly — he had to sit in the dayroom with its ever-present television. Rosenhan couldn't even safely send letters without interference. At the beginning of his stay, he was sending his secret observations about the hospital back home through the mail. He had developed a code to get his messages out — to make it look like gibberish (as if Rosenhan needed help with that since his handwriting accomplished this goal on its own), he skipped every other line and then looped to the top of the page to fill in the lines that had been skipped with new writing. When Rosenhan licked the envelope, the nurse, Mrs. Morrison, asked him not to seal it because the staff would have to read his letters before they were sent out. "Not everyone reads them," she reassured him. "Just the doctors and nurses." But when there was no administrative reaction to the content of his mail, he soon realized that no one gave a damn about what he wrote on the ward, so he stopped mailing letters altogether and just wrote in his diary out in the open for everyone to see.

Powerlessness. This is a word he repeats often in his notes. Patients lost many of their legal rights; movements were restricted; eating was confined to certain hours of the day, as were sleeping and watching television. The bathroom stink made its way to the dayroom as the urinals continued to overflow with human waste. The dormitory doors were locked. Rosenhan found one freedom that remained: his writing.

2/7/69
10:30 am

I've taken no pill but I'm exhausted, mainly from not sleeping last night. But also from boredom.

The dayroom drama unfolded in waves:
Over the monotonous yammering of the flickering television, two patients laughed so hard that they fell on the floor, appearing as if they lost control of their bodies.
A patient hit another patient.
Walter, one of the more disturbed patients, walked out of the bathroom nonchalantly

carrying balls of excrement up and down the hallway until an attendant finally noticed and made him wash up.

Sonny, one of the ward's troublemakers, hit a nurse and was dragged kicking and screaming into a lockdown room. Rosenhan almost missed the whole commotion, "so drugged was I from heat and the general torpor of the place," but everyone heard the sounds of Sonny pounding the hell out of his isolation room. "The walls here are plaster and no more — so there's reasonable chance he'll come through for a visit," Rosenhan joked. Gallows humor had already set in, after less than twenty-four hours on the ward.

But the joke was on him. It was time, a nurse alerted him, for his first meeting with his assigned psychiatrist, Dr. Robert Browning.

The interview lasted less than half an hour and mainly retrod the same topics that Dr. Bartlett had addressed in Rosenhan's intake interview. They discussed Rosenhan's financial difficulties, his "paranoid delusion" about a former advertising executive boss, and of course his vague auditory hallucinations.

THOUGHT LIFE AND MENTAL TREND:
Admits to ideas of reference, <u>delusions</u> of persecution evident regarding the friend he worked with in the advertising agency.

Auditory hallucinations present which have existed for the past six months and have gradually become more severe. They began as a lot of undifferentiated noise and followed by music, recently voices began, but they were not too clear. The voices said, "hallow and empty." "Also some sounds on that theme." They had become more severe this past month.

Dr. Browning found Rosenhan's speech "mildly constricted," meaning that he seemed to express a limited range of emotions. Outside the hospital Rosenhan would never have been accused of being unemotional, but inside, it seemed, an apprehensive look or detached tone was viewed as "mildly constrictive." On the outside people write; on the inside it's a sign of underlying illness. This is a vivid example of labeling theory in action — a phenomenon Rosenhan himself taught in his abnormal psychology class.

In 1946, Polish psychologist Solomon Asch studied the effect of certain "central" personality traits, such as "warm" or "cold" or "generous" and "ungenerous," descriptions that are so powerful they completely shape how we view others. There are few more powerful descriptors than "crazy" or "insane." In another later experiment, two psychologists played a recorded conversation between two men to clinicians. Half were told that the interviewee was a job applicant, the other half that he was a psychiatric patient. Those who thought they were listening to a job applicant deemed him fairly well adjusted and used terms like "realistic"; "unassertive"; "fairly sincere, enthusiastic, attractive"; "pleasant, easy manner of speaking"; and "responsible" to describe him. Those who believed he was a psychiatric patient used words like "tight, defensive"; "conflict over homosexuality"; "dependent, passive-aggressive"; "frightened"; "considerable hostility." Once words like *mental patient* or *schizophrenic* are affixed to you, there is little you can do or say that can make them disappear, especially when anything that doesn't support the doctor's conclusion is discarded for evidence that does.

How much of Rosenhan's diagnosis, "constricted speech," and "delusions of persecution" emerged from the expectation of how a mentally sick person *should* look

171

and act? I recognized so much of this. During my own time in the hospital, I remember a psychologist noting that I wasn't able to read or focus my eyes directly in front of me. It was only after I'd spent several weeks in the hospital that she realized my vision issues occurred because I had contacts lodged in my eyes. When I was deemed crazy, no one seemed concerned about my vision. My perceived craziness had colored everything else — even my eyesight.

This was a typical outcome of "the medical gaze," the dehumanization of patients first described by Michel Foucault in his 1963 book *The Birth of the Clinic: An Archaeology of Medical Perception.* Foucault wrote that this detached way of looking at illness emerged during the Enlightenment, as doctors learned more about the body, relying on empirical knowledge, rather than on magical thinking, to diagnose. Since then, clinicians had grown so reliant on these objective facts in the form of charts, percentages, and test results that they no longer *saw* their patients. Rosenhan's experience was a perfect example of such clinical blindness — the doctors read Rosenhan's chart but failed to see the patient standing right in front of them.

Beyond the perceived issues with his

speech, the doctor found Rosenhan to be an otherwise reasonably intelligent man who was sufficiently oriented to time and place. He could recall a series of eight digits forward and backward and could subtract from one hundred by sevens. When he asked Rosenhan to interpret a series of proverbs, the doctor was noticeably impressed. For the proverb "One man's meat is another man's poison," Rosenhan responded, almost without thinking: "Good for one, bad for another." For the proverb "A stitch in time saves nine," Rosenhan responded: "An ounce of prevention is worth a pound of cure." Touché. Then: "Do not cross your bridges before you come to them." Rosenhan's interpretation: "Don't try to anticipate a situation." How apropos.

Yet the doctor concluded that Rosenhan was suffering from schizophrenia, this time reducing the diagnosis to "residual type," defined as a person who has exhibited signs of schizophrenia but is no longer psychotic. This was a different diagnosis from the one that had landed him there just a day earlier: schizophrenia, schizoaffective type. Psychiatrists, steeped in the psychoanalytic tradition, shrugged off these differences as nonessential — you say potato, I say residual type.

173

■ ■ ■ ■

Rosenhan's clothes, which he'd now worn for twenty-four hours, smelled of the ward. Nothing upset him more than this indignity. He wanted his belongings, but every time he asked for the bag that had been confiscated during his intake, they refused him. It became an obsession. He found himself muttering under his breath about his lost clothing.

"Have my clothes come up yet?" he asked an attendant.

"What clothes?"

Rosenhan sighed. "I came into the hospital with some clothes and they were left downstairs to be marked. Could you call now?"

"No, they're probably closed. Will try if they don't come up at four."

"But they're more likely to be closed at four," Rosenhan said.

"We'll see," the attendant said. "Keep the faith."

During a shift change before going to bed, Rosenhan again asked for his bags.

"They came yesterday," a new attendant said, checking the label.

When Rosenhan made a face, he responded: "Well, he probably didn't see them under the desk."

Nurses' Note: 2/8/69 Very quiet. Taking
notes on other patients. No problems on
the unit.

While awaiting Mollie's daily visits, Rosenhan passed the time "whiling it away," which
he defined as "the daydreaming, the snooz-
ing, the coffee sipping, and the long inspec-
tions of space." Saturday was the dullest
day, when the ward was understaffed and
the psychiatrists and psychologists were
home with their families. He learned the
unofficial rules. Queue up when medications are dispensed (so you can spit them
out quickly in the bathroom with the other
patients); get cigarettes lit by other patients
instead of waiting to find a staff member;
arrive at the cafeteria quickly, as getting
there late meant missing out on the truly
edible items like bread, sugar, creamer, and
desserts. Another ward rule: The healthier
you were, the more the psychiatrists stayed
away. In other words, the saner you appeared, the more invisible you became.

Without grounds privileges, Rosenhan was
a literal prisoner. He managed to cheek the
pills — two milligrams of Stelazine, an anti-

175

psychotic; and twenty-five milligrams of Elavil, an antidepressant — but still he was groggy, drugged by the place itself. The blinds were open regardless of the sun's glare. The patients' discomfort didn't matter one iota to the nurses, who hardly left their cage (everyone was a prisoner there, it seemed). In Rosenhan's notes, he made rough estimations of their comings and goings, finding that they spent only half their time on a ward and a mere fraction of that interacting with patients. The staff existed in a different world — they ate separately, gossiped separately, and even used their own bathrooms, "almost as if the disorder that afflicts their charges is somehow catching," he would later write.

At one point a nurse in full view of twenty male patients opened the first five buttons of her uniform and adjusted her breasts. "No, she was not being seductive," Rosenhan wrote. "Just thoughtless."

Eventually Rosenhan spotted two newspapers on the ward for the first time — the local paper and a week-old *New York Times* dated January 31, 1969. Rosenhan snatched it up, desperate for something to distract him. He wrote in his notes:

"Where is today's paper?" I ask a nurse.

176

"Doesn't come until the afternoon mail." Which is to say that the paper has been coming everyday but the patients never see it.

He flipped through articles on the growing arms race against the Soviet Union and the launching of the Sentinel antiballistic-missile system. Nixon announced a plan to replace the draft with volunteers. Ads for Frank Sinatra Jr. playing at the Rainbow Grill at Rockefeller Center ran alongside news of renewed fighting in Laos.

After reading the paper, Rosenhan returned to his own writing.

"Would I have to be secretive? Hardly. One guy rocks, another leans, and I write."

The third day's diary entries are filled with musings on the hierarchy of the hospital, which he described as a pyramid structure with psychiatrists at the top, nurses just below them, and patients at the very bottom, of course. Skin color, he noted, also determined rank. Attendants, a notch above the patients, were almost all black. They were also paid the least, treated the worst, and had the most hand-to-hand contact with the patients. Rosenhan identified them as fellow "nether people."

"I'm Bob Harris." The sound jolted

Rosenhan back into the world of the day-room. The voice belonged to one of the attendants he had met his first day in. Harris offered his hand and Rosenhan shook it, delighted by the unexpected intimacy of the moment. No one here had yet greeted him this way; most didn't even lift their eyes. "I've been on the ward for six months now. You're new here?"

Rosenhan said he was. Harris told Rosenhan a bit about himself: He was struggling financially and working two jobs (the other at a gas station) to make ends meet to support his wife and three children. He planned to train as a nurse because the pay was much better than the fifty-five dollars a week he was making as an attendant.

The two chatted about the ward and its patients. "Now Jumbo, he's one I don't understand," Harris said. "He got no family so far as I can see, except an occasional friend that comes to visit, and he hasn't visited for months. He's got a very hot temper. Couple of months ago he just tore off at Harrington for no reason at all. I'd watch out for him."

Then there was Carroll: "With a name like that no wonder he's got troubles. I think he's been babied too much, even here on the ward. Mrs. Purdy really looks out for

178

him. Same for the kitchen staff. He always gets another dessert, you can be sure of that." Sam was in "because of homosexuality," and Peter "gets the largest Thorazine dose on the ward." Then Rosenhan's roommate shuffled by. "He's new. Probably been hospitalized before. Doesn't he just look like someone who's been in and out of hospitals since the war? Surprised he's not at a V.A. Hospital. They've got him in a room with those two kids, Drake and Foster. He won't notice it, but they're trouble. They're here on court orders and their lawyer has been in several times to see them. Drug rap."

Rosenhan nodded away, hoping that the conversation would continue, as it was the first real one he'd had since Mollie's visit the previous day. Harris moved on to the staff. The foreign residents weren't very good, except "a really good Cuban" named Dr. Herrera, he said.

After nearly an hour, Harris noticed the group of nurses in the cage waving him over. He excused himself, saying he'd be right back: "There's a lot more about this place."

Rosenhan felt a warm rush of gratitude. Perhaps this place wasn't so bad after all. This attendant had treated him like a person, not a leper. But as Rosenhan watched, he saw that the nurses were doubled over in laughter. They handed Harris a chart.

Could they be laughing at him? Was Rosenhan growing paranoid? What could be so funny about a middle-aged man with a family ending up in a psychiatric hospital?

Harris did not return to Rosenhan's table as promised. And when Rosenhan bumped into him later that same day, Harris's demeanor had clouded.

"Mr. Harris?"

"I'm busy now."

Rosenhan allowed himself to be brushed off — perhaps Harris was in a bad mood or something troubled him on the ward. But when he tried again later near the patients' bathroom, Harris still seemed irritated.

"Mr. Harris." Maybe he didn't hear. "Mr. Harris?"

"Didn't I say I was busy?" he snapped.

Normally Rosenhan wouldn't have taken such insolence without comment, but he couldn't muster up the reserves to defend himself. He was so distressed that he scribbled a quick note: "Even Harris' differentiated friendliness runs rapidly into friendly disdain."

Day Four

Nurses' Note: 2/9/69 Patient spends a lot of time by himself writing and watching TV

Each day seemed to yawn into the next, especially on that wintry Sunday with its skeleton crew. Harris, the only attendant on duty, continued to avoid Rosenhan. People walked the halls hunched over with blankets wrapped around their shoulders like depressed ghosts. Rosenhan joined the pantomime, pacing up and down the hallway with his own blanket and a blank expression. "The pacing, the sitting, eyes glued to the TV, was something that I, a sane man, came to do, often and for long periods of time. Not because I became crazy — at this writing, 72 hours after I came in, I still think I'm sane, although I can't guarantee my future — but because there is simply nothing else to do. How can I communicate the daily boredom, punctuated for me by my wife's daily visit but nothing for the others? The apparently psychotic behavior is not determined by psychosis at all — but by ennui."

Rosenhan choked down breakfast and returned to the drafty dayroom, where he fell back into uneasy slumber. He woke for lunch — "pink gloppy," a white sauce with

181

pale-pink things floating in it — prompting a diatribe in his notes from a man who prided himself (thanks to a mother who was an awful cook) on his ability to choke down just about anything. "The accounting department has obviously taken over the kitchen . . . Cook better, serve better foods, damn it and the 'proper balanced diet problem' will disappear!" This is all contained in his private writings; none of it was communicated aloud.

Rosenhan began to warm up to the patients, many of whom he initially expressed a "nameless terror" of. "Distance permits us to control the terror, to keep it from awareness — away!" he wrote. But as a patient, he could maintain only a fingernail hold on that distance. He asked around about grounds privileges, which led to the inevitable question, *How do you get out?* A patient named Bill summed it up: "You got to talk to the doc. Not in his office but on the floor. Ask him how he is. Make him feel good."

Make the *doctor* feel good? Who was running the asylum here? "Drs. exist to be conned," he wrote. He could hardly believe the level of manipulation that being a patient required, and how far one would go to avoid interacting with the system. Another patient, also named David, gave an example of how to play the game: "I might want to kill myself but I won't tell the psychiatrist, he'd keep me here," he said. "This way when I get out I can do what I want to." And yet another patient, Paul, who had been diagnosed with schizophrenia and had been in and out for years, had a similar perspective: "You've got to cooperate if you want to get out. Just cooperate. Don't assert your will."

Sunday 2/9/69
1:45 pm

I am depressed, sort of ready to cry. One tear jerking moment and I'd be flooding. Given my commitment to "being normal" on the ward, I can't account for my blues in terms of role enactment.

Later in the dayroom, after returning to the dining hall for dinner, he ran into the hostile Mr. Harris.

"Have you got a moment, Mr. Harris?" Rosenhan asked.

"Didn't I tell you to get away and quit bothering me?" Harris said.

Rosenhan watched himself flee from the interaction and "in doing so behaved like a patient." David Rosenhan, the professor, would never have allowed anyone — anyone! — to speak to him like that, but David Lurie, the patient, hung his head in shame. He went to the bathroom to splash water on his face and caught his image in the mirror. This time he didn't see just a haggard patient. He saw a middle-aged man in slacks and white button-down shirt (wrinkled, yes). The realization shook him out of his stupor: He looked like a professor, an academic, an intellectual. In much the same way that the judge had recognized Nellie Bly's ladylike demeanor, no beaten-up old Clarks or moth-eaten shirts could sufficiently mask Rosenhan's status. Harris, Rosenhan realized, must have mistaken him for a psychiatrist, and the intimate conversation emerged from Harris's desire to impress Rosenhan, whom he considered higher up in the pecking order. The illusion dissolved when the nurses broke the news. The look on Harris's face — total embarrassment — returned to Rosenhan and he felt vindicated. *He thought I was sane.* But the relief was fleeting.

Rosenhan begged for a phone call to check up on his family, but the nurses wouldn't budge: He didn't yet have phone privileges. These were doled out in stages

— first phone, then grounds, then day passes, and finally night passes until you were stepped down to one of the open Osmond-style buildings or released. Rosenhan still needed to prove that he could use the phone responsibly. "I then had the fantasy of kicking the door, trying to break it down." He imagined swaggering into their darkened cage. "You think I'm a real patient! I'm not. I'm sane. I faked my way into the hospital for a study I'm doing. In fact, I'm not David Lurie, I'm David Rosenhan, professor of psychology!"

But the fantasy always ended the same way, much as it did when Bly had tried in vain to convince doctors of her sanity: with the nurse asking, "Do you often think you're 'David Rosenhan'?"

Day Five
Nurses' Note: 2/10/69 Patient quite cooperative. Patient had visitors this PM. No complaints at this time.

Rosenhan was in a foul mood when he woke up on the fifth day to an attendant berating a patient for using the shower too long. "The blood rises," he wrote. When he stumbled to the bathroom and discovered that the door's handles had been unscrewed

185

the night before, destroying even the illusion of privacy, "the blood rises further." In the cafeteria on pancake day (which sounds far better than it was), Rosenhan asked the lunch ladies for some syrup. They directed him to an attendant who was eating by himself in the back of the room with the one maple syrup container.

Rosenhan asked the aide to pass him the syrup.

"There is none," the attendant responded. "You've got to use jelly." Rosenhan stared as the aide poured a river of the brown liquid onto his already syrup-logged pancakes.

Rosenhan was so angry he nearly blurted out: "Are we supposed to be blind?" But he stopped himself, recognizing in time that *anger, however justified, is here considered sick, disturbed.*

And he wanted out. The words of one patient stayed with him: "Don't tell them you're well. They won't believe you. Tell them you're still sick, but getting better. That's called insight, and they'll discharge you."

Back in the dayroom, he continued writing.

"What are you writing?" a fellow patient asked.

"A book."

"Why do you write so much?"

It wasn't the first time one of his peers noticed his constant writing. Another patient had asked him if he was penning an article about the place. Others had asked outright: "Are you an undercover journalist?" One psychiatrist seemed to have caught on, at one point commenting, "What are you doing, Mr. Lurie? Writing an exposé of us?" When Rosenhan asked him to repeat his question, the doctor waved it off. It was just a joke. Of course David Lurie wasn't writing an exposé. That would be crazy.

In the dayroom, Rosenhan witnessed a scene between Harrison, an attendant who had greeted Rosenhan with a razor his first morning in, and Tommy, an eighteen-year-old diagnosed with schizophrenia.

"I like you Mr. Harrison."
"Get over here."
Harrison pushes Tommy into his room.
"Where is your bed?"
"Please don't. I didn't do anything."
Harrison tosses Tommy onto the floor and pins him down, knee on arm and stomach. Tommy cries out and fights back. [Harrison] is now openly angry, throws Tommy

187

onto his bed, reaches under and appears to grab his balls.

A nurse interrupted the assault. She threatened to lock Tommy in solitary.

Tommy later struck a patient in the face, and this time the nurse did not hesitate to send him into an isolation room. He kicked and screamed and thrashed and yelled with such violence that it took two attendants and a nurse to push him inside. Rosenhan watched Tommy through the glass opening at the top door:

He began to break the walls, first with the bed and then with his bare hands. No one stopped him as he screamed and cried, his hands and even his face and arms bleeding from the torn plaster. No one administered a calming sedative. Rather, nurse, attendants, and patients watched through the little window that opened onto the isolation room, crowding each other for the pleasure of watching a nether person tear himself into bloody exhaustion.

Day Six
Nurses' Note: 2/11/69 Quiet and co-operative with no known complaints.

Spends a lot of time in dayroom watching
TV and writing

It must have been a nurse who led Rosenhan to the ward's conference room. Did he
lose his composure once he saw the ten or
so pairs of eyes — some of them no doubt
strangers — narrowing to take him in?
Certainly there were his two psychiatrists,
Dr. Bartlett and Dr. Browning, and the
ward's head nurse, but there must have
been unfamiliar faces, too, like the chief of
male services, the clinical director, and a
social worker or two, all there to make an
assessment.

These did not always go smoothly and
respectfully. In a case conference in 1967, a
patient admitted that he suffered from
syphilis and one of the doctors asked him if
he had sores on his penis. The man shook
his head, but the doctor ordered him to
drop his pants in front of the entire room.
No one questioned the doctor or thought
about what effect this might have on a
person who was already psychologically
fragile. The psychiatrist was king.

This was a new case conference — typically people on the ward had several. But
Rosenhan didn't want another meeting. He
wanted *out.* He took the advice that the

189

other patients had given — convince them with a narrative they would understand. He would say that he had hit rock bottom and Haverford Hospital had helped him climb out of it. Rosenhan explained that prior to his hospitalization, he had secured an interview with an advertising agency in Philadelphia. It was a big opportunity. It was time to leave.

The staff dismissed Rosenhan from the conference room so that they could discuss his case. They changed his diagnosis again, now to "acute paranoid schizophrenia, in partial remission," and granted him a day pass to attend the interview. They also recommended that his commitment run out, meaning that he would soon be free to leave. But they insisted that it was important for him to continue outpatient psychotherapy.

Day Seven

Meanwhile, the hospital decided that Rosenhan was now healthy enough to walk the property unaccompanied, and gave him grounds privileges ("in record time!" he wrote). He could join in ward activities, go on walks, and use the phone. Privileges allowed him access to the gym — where he "couldn't tell many of the patients from the staff," he wrote. That hollow dread he felt in the presence of "the other," the patient, was gone.

After gym, he joined the chattel outside the cafeteria waiting for the doors to open, pacing back and forth to pass the time.

"Nerves?" Faust, an attendant, asked.

"Bored, nothing to do."

Rosenhan's behaviors were self-fulfilling prophecies: He was crazy so he paced; he paced because he was crazy. Though there were many reasons for the pacing — sheer boredom, for one — the diagnosis shaded every interaction, every movement, and even every footstep.

Later that morning he overheard a conversation in the bathroom. One of the attendants was shaving a patient, who winced from the cold water and the feeling of the dull blade against his neck.

"Look, this may be cold, but it's the best we could do," the attendant said.

Rosenhan laughed. *This* is the best you could do?

Day Eight
Nurses' Note: 2/13/69 8:30 pm — Patient returns from temp visit. [Stated he had a nice time.]

The hospital released Rosenhan on a temporary visit to attend his "interview" — but I imagine he spent the day with Mollie and his children. No writing exists for this day in his notes or in his book.

Day Nine
Nurses' Note: 2/14/69 Patient is being discharged. custody of wife.

8:35 am

It's not so easy to leave.

Did he literally mean it wasn't so easy to get out or did he mean it was hard for him to summon the psychological distance necessary to move on? It's unclear. In his final notes on the ward, Rosenhan waxed poetic about the patients and his new friendships (whether this was authentic, exaggerated, or a function of the relief of leaving is hard to say): "Feel like I'm leaving friends behind. One develops a camaraderie of the afflicted, the cursed, and one's good fortunes feel like misfortunes."

By midday, Rosenhan's notes took on a more desperate tone. The doctor who was supposed to facilitate his discharge was late and there was a chance he would not arrive in time for Rosenhan to be released before the start of the weekend, at which point he'd be trapped there for another three mornings. Rosenhan smoked and smoked and smoked, trying to keep his nerves in check for fear that any sign of unease or aggression might lead to a renewed commitment.

And then, as if in a movie, Dr. Myron Kaplan arrived at zero hour. After finding Rosenhan competent to drive and "handle money," Dr. Kaplan released him to the care of his wife, out into the wintry world beyond the hospital. Dr. Kaplan recommended that he seek outpatient and "chemotherapy treatment" (a now outdated term for psycho-pharmacological medicine), leaving Rosenhan with a diagnosis, a prescription, and little else.

The patient was advised of the desirability of continuing out-patient psychotherapy, and he appeared to agree. However, he was somewhat ambivalent and undecided as to whether he could afford private out-patient therapy, or whether he would have to resort to a low fee therapist, or whether he would have to go up to a clinic for therapy. The patient was given a list of several clinics as well as the knowledge that he could consult the out-patient clinic

193

at Haverford State Hospital for a list of low fee psychiatrists, and he said that he would make a decision within the next few weeks. The treatment was chemotherapy and individual psychotherapy. The patient was discharged on 2-14-69, and he is competent to drive and handle money. Recommendation is that he continue psychotherapy on a out-patient basis. The diagnosis, therefore, is accute paranoid schizophrenia in remission.

Myron J. Kaplan, D.O.

You will notice that the doctor did not say Lurie was cured — no one was "cured" of a mental illness — but instead that he was in remission, much as cancer is during the beginning stages of recovery. The sickness could always relapse, and the threat of reoccurrence would remain with you like a sweat stain you couldn't quite scrub out.

Around the time of Rosenhan's first admission, researchers were studying the stigma of mental illness diagnoses. Stigma — in ancient Greece the word referred to a mark placed on slaves as a sign of their diminished status — created a sort of self-fulfilling prophecy that came externally (from the world around you) and internally (from your own feelings of shame). As

Rosenhan wrote in his paper: "A psychiatric label has a life and an influence of its own. Once the impression has been formed that the patient is schizophrenic, the expectation is that he will continue to be schizophrenic . . . The label endures beyond discharge, with the unconfirmed expectation that he will behave as a schizophrenic again."

This touches not only the patient but also the people around him. Study after study — from Rosenhan's time to today — has confirmed that people hold mostly negative views about people with serious mental illness. They are often viewed as more violent, dangerous, and untrustworthy. Three years after Rosenhan's stay at Haverford, in 1972, Tom Eagleton, a US senator running for vice president, lost his spot on the Democratic ticket when the public learned of his prior psychiatric hospitalizations for depression. With the Cold War raging, the question became: Do you really want this guy even near "the button"? It didn't matter that these hospitalizations had happened years earlier and that he had, by all estimations, recovered — once labeled, he, and others like him, would always be sick and would never be fully capable again.

I wish I knew how sweet the homecoming

was for Rosenhan and his family. I wish I could interview Mollie and hear her perspective. I wish I could see what he looked like, hear how he sounded. Was he tired? Were his clothes rumpled? Did he look like a different man? If I could, I would crack open their heads and pluck out the memories. Did he think of his brother during his hospitalization? Did he reframe some of his own behaviors in light of his new diagnosis? Did it frighten him to realize how easy it was to wear the garb of a so-called schizophrenic? Did his days on the ward touch on some paranoia, some part of him that felt unworthy? How many truths had his doctors chanced upon on the way to a gross misjudgment?

His research assistant Bea Patterson told me that Rosenhan seemed "quite shook" when he returned. "You could tell he felt whatever had happened to him [in the hospital] affected him deeply," she said. "He was quieter, more reserved." His abnormal psychology seminar students, some of whom I have interviewed, told me that when he returned from the hospital his mood had darkened. He seemed humbled. One student recalled that he looked distressed, worn out, somewhat older than before. The students begged to hear more, but he refused to discuss it. One thing was clear: They were not going to continue the experiment. It was over. Done.

The story could have ended here, an upsetting episode in the life of a professor who took on a difficult and painful role to protect his students. The study could so easily have remained a *what-if* — his notes would have likely been lost, his diary filed away, the experience reduced to an interesting footnote in Rosenhan's life. But it didn't.

Instead, sometime between the end of "David Lurie's" hospitalization in February 1969 and the first finished draft of his article "On Being Sane in Insane Places" in 1972, this single experience morphed from a teaching experiment into something much larger, as seven other volunteers joined — even though Rosenhan had declared that it was too dangerous — what would eventually become the study. They willingly subjected themselves to the same indignities Rosenhan had just survived, and in the process cemented Rosenhan's legacy in the history of psychiatry.

For as traumatizing as his days there had been, Rosenhan must have understood the value of his insight into life on the ward, and the importance of getting the "normal" world to finally pay attention. He needed them to listen in a way they hadn't listened to Nellie Bly, to Dorothea Dix, to Ken Kesey, or to any of the brave others who had gone before him. In order to bring attention to the state-sponsored travesties going on around them, he'd need more — more data, more hospitals, and more people to go undercover. He had to create an account that could not be dismissed. It needed to be solid, quantifiable. It needed to be *scientific.*

■ ■ ■ ■

PART THREE

■ ■ ■ ■

People ask, How did you get in there? What they really want to know is if they are likely to end up in there as well. I can't answer the real question. All I can tell them is, It's easy.

— Susanna Kaysen, *Girl, Interrupted*

11
GETTING IN

There is no question that "David Lurie" was, in fact, Rosenhan himself. But what about the others? They were not his Swarthmore abnormal psychology students, who had inspired the study. Who were they, then, and how did he find them? Why had they so selflessly decided to help Rosenhan in his quest to bring light to these dark corners? How would I find them now?

In Rosenhan's private writings, there was no insight to be found regarding how these people felt about their contribution to the history of medicine. Had it changed them as it did him? His unpublished manuscript gives only the sparsest of clues with no specifics about locations or time frames:

Chapter Three: Getting In

With the students out of the project, the entire study might have terminated for lack of manpower, were it not for an accidental encounter that occurred three months later. I was attending the meeting of the Society for Research in Child Development. It had been a long hard day full of heavy research discussions and disputes. A number of us were unwinding over dinner and I began to describe some of my experiences in a psychiatric hospital. Afterwards, a couple who had been at dinner whom I had not previously met, came over and introduced themselves. We talked deep into the night about psychiatric hospitals and psychiatric care.

This was the couple he called John and Sara Beasley, recent retirees who had each logged many years in the mental health field, John as a clinical psychiatrist and Sara as an educational psychologist. The prior six months, they had traveled and read, thoroughly enjoying their retirement, all while keeping up on developments in their fields, which is how they ended up on March 29, 1969, at Rosenhan's lecture on altruism in children in Santa Monica, California. The three hit it off. Of John, Rosenhan wrote: "It was his thoughtfulness that was especially striking, quite as if he had used the past six months of retirement to ponder the nature of psychiatry as he and others had practiced it." Of Sara, he wrote, "I should have been delighted to confide my children's school problems to her. She seemed to combine deep knowledge of children's (and parents') problems with a firm optimism that they could be solved."

Rosenhan met John and Sara two days later for dinner. "John was particularly struck by the symptoms that I had used. They reminded him of a question he had asked himself quite frequently: how well was he able to predict a patient's behavior and in particular how much of what he thought he saw in patients was really there. Moreover, he was quite interested in obtaining a firsthand picture of treatment," Rosenhan wrote. By the end of dinner, John had decided he would like to try Rosenhan's experiment for himself. Rosenhan coached John about the "thud, empty, hollow" symptoms and taught him how to cheek pills. "The procedure was simple, but it involved some gall," Rosenhan wrote. "After placing the pill on your tongue, you needed to flip it underneath and then drink the water that was given, all the while looking the nurse straight in the eye." They came up with an occupation: John would be a retired farmer (for he lived on a defunct farm and was familiar enough with the work to fake it). They talked about how to get in, how to take notes, and the importance of having daily visitors.

Six months later, in October 1969, John called Rosenhan with news: He had just left Carter State after having spent twenty days there with a schizophrenia diagnosis. John's wife, Sara, was currently undercover, too, and John's sister, known as Martha Coates, planned to go in. Rosenhan's school project was suddenly multiplying like bacteria left overnight in a petri dish.

Rosenhan recounted bits of John's, Sara's, and Martha's hospitalizations in his unpublished book, quoting, he wrote, from their diaries and notes. John described the absurdist drama of hopscotching beds his first night. In the morning, he woke to a strange man sitting on the edge of his bed. "Bearded and burly, the combination of size and gentleness scared the daylights out of me," John wrote. "He told me quietly 'It's time to get up.' All the other patients were still asleep. I could see that the ward wasn't up yet. But he insisted that I get up, and moved the covers off me. It was Kafkaesque."

Sara had admitted herself to Westerly County Hospital, a smaller teaching hospital close to her home. Though Rosenhan acknowledged that he "doubted" that Sara would join the study, he didn't comment on why she ultimately did decide to, especially after her husband's experience was so distressing.

"I don't know what's troubling me," she wrote in abbreviated shorthand, according to Rosenhan's book. "I've never felt so uncomfortable with psychotic people before. There's no reason for it." She tried to make sense of her fear: "Maybe it's because I lied my way in . . . Maybe it's that I can't tell what the patients will do next? But they seem to be doing very little. Most of them are drugged . . . So what if they check my bed once or twice? I can't seem to get under control. Maybe I should swallow that medication. Careful now." The tension abated after the second day, leaving almost as suddenly as it appeared. "I feel much better now," she wrote on the morning of her third day. "I don't know why, I hope it lasts." Sara spent a total of eighteen days hospitalized and was released with the same diagnosis: paranoid schizophrenia in remission.

Despite this unsettling experience, John was more devoted than Rosenhan himself had been, and decided one time through the wards wasn't enough. He readmitted himself, this time to another, larger hospital called Mountain View, spending two more weeks institutionalized and once again diagnosed with schizophrenia. Before, he was focused on maintaining his charade; this time, Rosenhan wrote, he wanted to focus more on the patients and "evaluate their distress before it became masked by medications."

John's sister, Martha, now the fourth pseudopatient, volunteered for what had grown into a family game of chicken. (What sort of family, one can't help but wonder, engages in this kind of brinksmanship for fun, or even for science? I was desperate to know more.) Martha, a recent widow and a housewife with no professional experience with mental illness, Rosenhan wrote, had a personal connection with the mission. Her son had struggled for years with drug addiction and had spent time in and out of psychiatric facilities. She expressed "some wonderment about what his experiences were like" and decided to re-create them for herself. Martha was also diagnosed with paranoid schizophrenia and was released two weeks later with the illness "in remission," now the fourth patient in a row to receive the same outcome. The symptoms that Rosenhan had devised as a joke in his Swarthmore class, "thud, empty, hollow," seemed to have become a shortcut for doctors to a diagnosis of schizophrenia.

Rosenhan didn't report many details in the interim about signing up the other pseudopatients for the study, but he does write that six months after John Beasley's first hospitalization, a "famous abstract artist" — successful enough for major museums across the country to feature her work — named Laura Martin, the fifth pseudopatient, exhibited the same "thud, empty, hollow" auditory hallucinations and was admitted to the study's only private psychiatric hospital. Rosenhan named it William Walker Clinic, and described it as one of "the top five [hospitals] in the country." Like the other pseudopatients, Laura had no trouble getting in; her problem, even more than the others, was getting out. Laura was released against medical advice (the hospital wanted to keep her longer) after fifty-two days with the diagnosis of manic depression, the first pseudopatient to receive a different diagnosis than schizophrenia, which is telling because manic depression has more favorable outcomes. Could it be that her perceived social class in the context of a fancy private institution made her seem *less sick*?[1]

1. Yes, most likely, according to studies on social

Laura's husband, Bob, was the next to go in. He changed his occupation from pediatrician to lab technician and entered Stevenson State, an "otherwise unimpressive" psychiatric hospital. Twenty-six minutes into his intake, his psychiatrist diagnosed him with "schizophrenia, paranoid type" — the fifth such diagnosis. Becoming a patient was torture for the doctor. "The hamburger was so coated with grease that it looked and felt like slimy shellac. The potatoes were watery . . . I don't know how the patients eat this shit. I can't," Bob wrote. After seventy-two hours, Bob stopped eating cooked foods — only bread, butter, coffee and tea, and the occasional fruit. "I've never seen such lousy food in any hospital . . . I'm afraid everything is buggy," Bob wrote, according to Rosenhan. It got so bad that Laura and other visitors would sneak in food, like sandwiches and Oreos. Bob squirreled away the most disturbing parts of his meals — chunks of gray meat, unappetizing sauce —

---

class and diagnosis that date back fifty years. Older studies show that people with higher socioeconomic status were more likely to be diagnosed with manic depression (or bipolar disorder) than the general population. But more recent studies, however, have shown an opposite correlation.

in napkins just to show off how gruesome the fare was to his visitors. Rosenhan wrote about Bob in his unpublished book: "We ourselves were seriously concerned about his 'symptom.' Bob had not previously manifested any finickiness regarding food and was indeed regarded by some friends as omnivorous. His concerns about cleanliness in preparation, about the possibility of disease, his occasional comment about 'poison' had us sufficiently worried that had he not been discharged when he was, we would have removed him from the hospital." Bob was released on his nineteenth day and carried the label "schizophrenia, paranoid type in remission," but not one medical note addressed his one very real symptom: his refusal to eat. He left the hospital "hungry, sort of depressed, but smarter for it all."

Thanks to John, Laura, and the others, the data was pouring in. By the fall of 1970, Stanford had recruited Rosenhan to the campus as a visiting professor, based in large part on the reputation he was developing as the creator of this ingenious but still-unpublished study. He had lectured twice about his own experience, which he had titled "Odyssey into Lunacy: Adventures of a Pseudopatient in a Psychiatric Hospital."

In a letter to one colleague, he wrote, "With all due apologies for immodesty, the data are increasingly interesting." Others agreed. A *Psychology Today* editor wrote him a personal note to inquire about publishing the findings. Whispers of his work reached Harvard, which sent out feelers to him. Chairman George W. Goethals wrote: "There was further agreement that if this research 'took off' this would be a major contribution to American psychology."

During the wild summer of 1970, as the world was hypnotized by the murder trial of a group of drugged-out hippies and their mastermind, Charles Manson, Rosenhan headed west. He loaded up his VW and drove his young family to California, taking the scenic northern route. "The country is a hell of a lot more beautiful than most of what I've seen in Europe," he wrote to a friend. "Not only deep blue, but emerald green glacier fed lakes, that were symphonies in silence and isolation." Though their camera broke halfway and his daughter, Nina, caught chicken pox, Rosenhan described the road trip as magical. The urbanite couldn't get over Iowa: "I simply couldn't believe all that fertile land, and was totally taken by the rolling farms and decency of the Midwest. I could teach at Iowa, though it might cost me a spouse."

When he reached Palo Alto, any fantasies of rural life disappeared. "We've really lucked out here," he wrote in a letter to a former colleague at Swarthmore. "Palo Alto is a great place to live: civilized, urbane, with one hell of a lot doing." The view from his ranch house in the "Prof Hill" section near Stanford was magnificent, especially when the fog cleared to reveal the foothills of the Santa Cruz Mountains. Eight-year-old Nina touchingly told her father how "lucky we were to be here." Mollie tended to her new vegetable garden, picking pomegranates and planting a Meyer lemon tree, while Jack helped his father trim the hedges. Rosenhan soon traded in his VW for a gunmetal-gray 1957 Mercedes 190SL with red leather interior, his boyhood dream car. He was fond of the phrase "The coldest winter I ever spent was August in San Francisco" — an adaptation of a quote misattributed to Mark Twain — and used it to temper the glow of his happiness when sending notes to his colleagues out east. Despite making a pact with Swarthmore to return, he never did. A year after his arrival at Stanford, he was made a professor with a joint appointment in psychology and law. To Rosenhan, Palo Alto — the sunshine, the lush gardens, the Meyer lemon trees — must have felt like the academic land of milk and honey. He would spend the rest of his life tucked in the cradle of Silicon Valley.

Stanford University had already set its sights on establishing a world-renowned psychology faculty and devoted ample funds to make this a reality, recruiting some of the best and brightest minds. As a show of its newfound importance, the Department of Psychology moved to Jordan Hall, right at the center of the Quad, the same summer that Rosenhan arrived. Bold-faced names included child psychologist Eleanor Maccoby, a powerhouse who pioneered research in the study of sex differences and gender development; cognitive psychologist Amos Tversky, whose later work with Daniel Kahneman on cognitive bias and risk would fundamentally challenge the fields of economics, philosophy, business, and medicine; Walter Mischel, whose work *Personality and Assessment* shook up psychology by arguing that personality is not fixed; and of course the great Lee Ross, who set me off on this expedition.

"It was probably one of the most exciting academic places to be during that era," said Daryl Bem, who originated the "self-perception theory" of attitude formation, or when attitudes are formed by observing one's own behavior (say you're always in a bad mood when a friend visits you; maybe you'll conclude that you really don't like her). Bem worked at Stanford with his wife, Sandra Bem, famous for her work on gender and identity. "Everyone was intensely interested in their research. There's an old Jewish saying that there are only two admissible answers you can give if asked, 'What are you doing?' And the two answers are: 'I am studying the Torah,' and the other answer is, 'I am not studying the Torah,' " he said. "That's exactly how Stanford professors felt about their research. Either they were doing their research or they weren't." It was the only thing that mattered.

There was another perk to the move. As he would explain in his book, "one of the main motivations" for taking the new job was to "continue the hospital study." Stanford University offered him something Swarthmore College didn't: access to graduate students. By this time — seven pseudopatients in — he knew he was onto something huge: "The ease with which we were able to gain admission into psychiatric hospitals and remain there undetected was beginning to raise a question in my and my colleagues' minds . . . . Could it not have been the luck of the draw that got us admitted by the less talented members of these hospitals?"

He needed more data, which meant more willing volunteers.

Rosenhan spoke of one graduate student, Bill Dixon, a red-bearded Texan whom he described as prodigiously normal. Bill enthusiastically signed on to the study and, sure enough, spent seven days at Alma State Hospital with a diagnosis of schizophrenia.

It's unclear exactly when or how Rosenhan recruited Carl Wendt, pseudopatient #7, a businessman-turned-psychologist who had recently finished his PhD and planned to practice clinical psychology in a psychiatric setting. His interest in being a pseudopatient came from a desire to acquire firsthand knowledge. "Much as it is common practice to require of potential psychotherapists to undergo treatment themselves," Rosenhan wrote, "it seemed to make sense to Carl that he see what hospitalization is all about himself, before recommending it to patients." Carl's involvement lasted much longer than any of the other participants'. He would spend a total of seventy-six days locked away.

Carl's first hospitalization, at Memorial County, was the hardest. A psych interview, which lasted a mere twenty minutes, embarrassed the newly minted clinical psychologist. A bored shrink peppered him with questions in the following order: "What did you eat for breakfast? Have you ever wanted to murder your father? Did you grow up on a farm? Did you ever have sex with animals? Do you often feel that people are after you?" Carl recognized these questions from the Minnesota Multiphasic Personality Inventory, a paper-and-pencil psychological test used to assess patterns in thoughts or behaviors outside the norm; its revision is used today in everything from screening job candidates to legal proceedings.

Carl spent his first night in the middle of an open dormitory, crowded with patients and their bodily noises. In a scene that sounds similar to John's first night, Carl hunkered down only to find that a massive man had joined him under the covers and fallen into a deep sleep. An attendant moved Carl to the sleeping man's bed, which they found had been soiled. The only open bed (or what passed for one) was the plastic settee in the dayroom that separated two large dorms. Carl covered himself with a blanket and placed his hands over his ears to drown out the grunts, screams, and laughter echoing in the dayroom. He didn't sleep that night.

Per Rosenhan's notes, the next day Carl wrote in his diary: "I must be awfully tired. The place seems full of zombies."

By the third day, he wrote just two sentences: "I'm like a stone. I have never felt so inert."

Carl spent thirteen days at Memorial County before he left the hospital against medical advice with the diagnosis of paranoid schizophrenia in remission.

Once he was released, the depression lifted, and Carl (like John) volunteered to admit himself to Rice State, where he was released after thirty-one days with the same diagnosis. He admitted himself yet again at Godwin State, where he stayed for nineteen days. He went undercover a fourth time at Montadero Hospital, but this time Carl's enthusiasm for the study began to worry Rosenhan.

"Bizarre as it may seem, I was concerned that this unpleasant experience might nevertheless become somewhat addictive for him. For the purpose he had begun with, he had surely learned enough about psychiatric hospitals, at least for the moment," Rosenhan wrote. Thirteen days later, Carl was released, yet again with "paranoid schizo-

216

phrenia, in remission."

It was also Carl Wendt who would, perhaps entirely by accident, deliver the study's pièce de résistance. His fifth and final hospitalization was the most important of all — because it never happened.

This startling piece of the study grew out of a challenge from one of Rosenhan's colleagues. Unconvinced that Rosenhan hadn't simply been lucky in cherry-picking less prestigious or understaffed institutions, an unnamed doctor offered his own hospital — "an excellent treatment and research facility" — to Rosenhan. (The rumor is that this was the Stanford University hospital, but I have not been able to confirm this.) The psychiatrists, nurses, and social workers were confident that it "would never happen here" — and if it did, "they would soon detect it."

It was agreed that Rosenhan would address the staff at their weekly meeting, and alert them that in the next three months one or more pseudopatients would try to infiltrate the hospital. The orders were for the staff to document each admitted patient and rate how likely it was that the patient was an impostor:

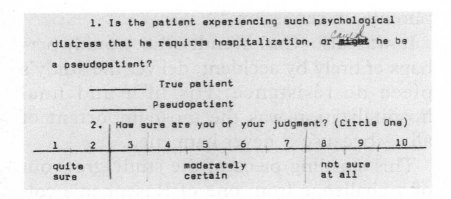

1. Is the patient experiencing such psychological distress that he requires hospitalization, or ~~might~~ *could* he be a pseudopatient?

```
_____  True patient

_____  Pseudopatient
```

2. How sure are you of your judgment? (Circle One)

1	2	3	4	5	6	7	8	9	10
quite sure				moderately certain			not sure at all		

Rosenhan arranged for Carl to go undercover at the unnamed hospital, but when the day approached Carl was waylaid by the flu. Rosenhan didn't replace him. Maybe Rosenhan forgot, maybe it was planned, he doesn't say — intentional or not, in the end Rosenhan sent no one. The result is, in my opinion, the most damning part of the study. Over the next three months, the staff kept an eagle eye out for any pseudopatients attempting admittance to their hospital. Of 193 new patients who arrived in that time period, staff members labeled 41 with high certainty as probable pseudopatients; psychiatrists, for their part, identified 23 likely pseudopatients. This miscalculation proved once again that we react to the world by what we are primed to see. In this case, these medical professionals were primed to see pseudopatients, so that is what they saw.

The *New York Times* interviewed Rosenhan about this ploy, which many newspapers focused on in the aftermath of the publicity craze the study created.

"Were the patients sane or not?" the reporter asked.

"We will never know," Rosenhan responded. "But of one thing we may be certain: any diagnostic process that yields so readily to massive errors of this sort cannot be a very reliable one."

Rosenhan had struck gold. And now it was time to share that gold with the world.

12
. . . And Only the Insane Knew Who Was Sane

And share it he did — in the splashiest way possible. Rosenhan landed a spot in one of the world's most respected generalist journals, one that mainlines directly into the veins of the insatiable mass media. So how did he accomplish this coup? I don't know for sure, but I have a suspicion that Rosenhan caught the ear of William D. Garvey, *Science*'s psychology editor, at an American Psychological Association conference that they both attended in 1970. However it came to be, Rosenhan then submitted his paper to famed *Science* editor Philip Abelson, a superstar researcher (co-discoverer of the chemical element neptunium, whose work on uranium contributed to the creation of the atomic bomb), in August 1972, summarizing his findings as follows: "The article presents experimental data on our inability to distinguish sane from insane people in psychiatric institutions. It also briefly describes the experience of psychiatric hospitalization as observed by pseudopatients."

When Rosenhan's study hit *Science* in January 1973, fan letters flowed into his Stanford office from around the world. Psychiatrists from Camarillo State Hospital down Highway 101 wrote to add their own anecdotal evidence confirming Rosenhan's thesis about the ineffectiveness of psychiatric diagnosis; Robin Winkler, a psychologist from Australia, shared some of the data he had gathered doing his own pseudopatient-centered research down under; Thomas Szasz offered congratulations, as did Abraham Luchins, one of the country's most important Gestalt psychologists, who pioneered the use of group therapy. Students wrote in asking to join his study. Former and current patients begged Rosenhan to prove that they, too, were sane people in an insane place and *please could you get us out?* "I read your article published Wednesday, March 1973 in the Huntington, West Virginia newspaper entitled: 'Eight Wonder Who Can Tell The Sane From the Insane.' I am number 9." The letters were sent from all around the country from the known and the unknown, the famous and the infamous, including one that read "My name is Carl L.

Harp. I am charged with murder and assault here in Seattle, Washington. 'Bellevue Sniper.' I am innocent." Another: "Dear Dr. David Rosenhan, I am a 29 year old black, militant, social democrat . . . Those state hospitals are no more than concentration camps . . . Why can not the wealthiest country in the world take proper care of its mentally ill?"

Rosenhan personally responded to almost every letter. He engaged — sometimes with wit, sometimes with professorial authority, but always with interest and compassion. In response to the "black, militant, social democrat," for example, he wrote: "I couldn't help but wonder whether someone who is black and militant and who had moved from a small town in Louisiana to Cambridge, Massachusetts might not be expected to experience some stress. And I wondered further whether the stress might not be misdiagnosed as schizophrenia. Obviously I can't tell — one doesn't diagnose long distance. Regardless of diagnosis, it seems to me that you've been through one hell of a lot."

(Most of his letters, it should be noted, were typewritten. He seemed to be aware of the impenetrable nature of his distinct handwriting, as he explained in a letter to his former student Pauline Lord: "I hope you forgive me — I never handwrite . . . I still write in hieroglyphics, but without a Rosetta Stone.")

Rosenhan capitalized on this fame, lecturing widely about "The Horrors of Your Local Mental Hospital." People were riveted. I can imagine the sound of Rosenhan's resonant voice echoing out over the crowd as he swaggered around onstage, high on the life of a man-in-demand as the world begged him to visit their institutions, their fundraisers, their conferences, their causes, because everyone wanted a piece of him, because he had proven what everyone already suspected was true.

The media co-signed. As of my count, which is by no means definitive, seventy local and national newspapers, in addition to television and radio shows, covered the study. Some, like the *Los Angeles Times,* ran it straight: "Eight Feign Insanity, Report on 12 Hospitals." Others used it to anchor editorials, like the *Independent Record* in Helena, Montana, which posed the question: "Can Doctors Distinguish the Sane from the Insane?" Others took a more creative approach: The *Burlington Free Press* headlined its piece: " 'Mania,' 'Schizo' Labels Cause Wrangle." The *Palm Beach*

Post used: ". . . And Only the Insane Knew Who Was Sane." Immediately after its publication, two publishers approached Rosenhan about turning his study into a book. He signed on with an editor at Doubleday in May 1973. By the following year he had finished eight chapters — a good chunk of the book for which the publisher would, almost a decade later, be forced to sue him when he never delivered the manuscript.

The study smashed through the one-way mirror separating the layperson from psychiatric jargon and its judgment. Young upstart lawyers who had read Rosenhan's study would trot it out in court to undermine the validity of a psychiatrist's expertise on the stand. A year before Rosenhan's study, ACLU lawyer Bruce Ennis had indicted the whole field of psychiatry by calling it an "enterprise" that treated patients as criminals in *Prisoners of Psychiatry*. Ennis and others maintained that psychiatrists were no more reliable than flipping coins — and that they "should not be permitted to testify as expert witnesses." In the wake of the study's publication, judges increasingly overruled expert testimony by psychiatrists, especially when the doctors recommended psychiatric commitment.

During a time when the president was insisting, "I am not a crook!" Americans could understand a study like this — one so sensational, yet seamlessly commonsensical — that provided a scientific basis for what so many of us had already experienced: The world was topsy-turvy and no one could prove who was on top and who was on the bottom.

Today the various factions that write about psychiatry agree about very few things, but do concede this: Rosenhan's study had an overwhelming effect not only on public opinion but also on the way that the field saw itself.

"When the Rosenhan study was initiated it was right around the time that the Emperor's New Clothes were about to come off," Columbia psychiatrist Dr. Jeffrey Lieberman, the author of *Shrinks,* told me in an interview. "Rosenhan, I think dramatically and very effectively, pointed out glaring weaknesses in our knowledge base and our methods for making psychiatric diagnoses, and exposed it as fallible."

"Rosenhan's study was akin to proving that American psychiatry had no clothes. It was evidence that American psychiatry was diagnosing schizophrenia in a willy-nilly, frivolous manner," wrote medical journalist

225

Robert Whitaker in *Mad in America.*

"It was a landmark study that shook us all — it created a crisis of confidence," said Allen Frances, the architect of the *DSM-IV.*

"The most celebrated psychological experiment of the era . . . [showed] that psychiatry — like mental illness — was a myth . . . as evidence accumulated that there simply was no there there, as Gertrude Stein might have said," wrote *Madness Is Civilization* author Michael Staub.

If psychiatry could get its bread-and-butter diagnoses wrong, what else was it wrong about? A lot, it turned out. It was no coincidence that at the same time that Rosenhan's study made waves, the field was facing another reckoning in the form of "the homosexuality problem."

Being gay then was considered a mental illness — more specifically a form of "sociopathic personality disorder," according to the *DSM-I.* (When Rosenhan arrived at Stanford, there was a joke going around about the professor who asked if the department would hire a gay professor. The answer: "You could be an ax murderer as long as you did that on your own time.") Not only did gay Americans risk getting arrested (sodomy between consenting adults, for example, was still illegal in forty-nine states as of 1969) or losing their jobs; they could also be committed to a mental hospital. Psychoanalysts had given this belief a foundation. They claimed that homosexuality was pathological and emerged from unhealthy family relationships. In a widely read layman's book, psychoanalyst Edmund Bergler charmingly asserted: "Homosexuals are essentially disagreeable people, regardless of their pleasant or unpleasant outward manner . . . [their] shell is a mixture of superciliousness, fake aggression, and whimpering." (He added: "I have no bias against homosexuals; for me they are sick people requiring medical help.") Before he became president, Ronald Reagan said, "We can debate what is an illness or whether it is an illness or not, but I happen to subscribe to the belief that it" — meaning homosexuality — "is a tragic illness, a neurosis the same as other neuroses."

Some psychiatrists started to direct a more "biological" approach to "treating" homosexuality. "Homosexuality is in fact a mental illness which has reached epidemiological proportions," said psychoanalyst Charles Socarides, an infamous practitioner of conversion therapy, which tried to "cure" gay people with analysis. Robert Galbraith

Heath, of Tulane's electrical brain stimulation program, was one such practitioner of bodily "cures" for the "homosexuality problem." In 1970, Heath implanted electrodes onto the brain of patient B-19, a gay man, and subjected him to rounds of electrical stimulations as he watched heterosexual pornographic movies. According to Heath's records, the patient reported "continuous growing interest in women" to the point that he wanted to consummate a sexual relationship with one. Heath obliged and brought a twenty-one-year-old prostitute into the lab. Despite the inhospitable surroundings, B-19 "ejaculated" and left the sickening experiment "cured," at least according to Heath.

When news of the story reached the public, the Medical Committee for Human Rights protested at one of Heath's events and a local journalist published a long account of Heath's work titled "The Mysterious Experiments of Dr. Heath: In Which We Wonder Who Is Crazy and Who Is Sane," a clear reference to Rosenhan's study.

Gay rights groups had already started fighting back. The same year that Rosenhan started his study, police officers staged a raid on a gay bar in the West Village, inscribing the name Stonewall in the history books and galvanizing the gay rights movement.

But to win the larger civil rights battle, gay men and women had to force doctors to stop labeling their sexual preference a medical condition.

In May 1970, gay activists infiltrated the American Psychiatric Association's conference in — of all places — San Francisco and "shrinked the headshrinkers," disrupting seminars and forming a human chain around the facility. "This lack of discipline is disgusting," said psychiatrist Leo Alexander at the meeting. He diagnosed the problem of one of the protesters. "She's a paranoid fool," the doctor said, "and a stupid bitch." The optics weren't great for psychiatry. A year later at the APA conference in DC, Dr. Frank Kameny, a gay rights advocate who had lost his job as an astronomer when the US Army's Map Service learned about his sexual orientation, grabbed the microphone and yelled: "Psychiatry is the enemy incarnate. Psychiatry has waged a relentless war of extermination against us. You may take this as a declaration of war against you."

Psychiatrists addressed these issues head-on at the 1972 APA meeting in Dallas with a panel with the tone-deaf title "Psychiatry: Friend or Foe to Homosexuals?"

One panelist was John Fryer,[1] a young psychiatrist who lost several jobs when employers became aware of his sexual orientation. Fryer agreed to join the panel on one condition: He would remain anonymous. Fryer went to Uniforms and Costumes by Pierre on Walnut Street in Philadelphia and bought a sagging, flesh-colored mask and a black curly wig. He paired these with a baggy tuxedo with velvet lapels and a velvet bow tie, making an unsettling figure sitting on the APA panel. When he spoke, a special microphone distorted his voice as he read from his notes:

THANK YOU, DR. ROBINSON

I AM A HOMOSEXUAL. I AM A PSYCHIATRIST.

1. Fryer would cross paths with Rosenhan in 1973 when he arranged for Rosenhan to attend a symposium at his hospital, Norristown State Hospital near Philadelphia, on the topic of "The Rights of the Mental Patient." Another guest? Dr. Bartlett. During that same visit, Fryer also arranged for Rosenhan to go undercover as a pseudopatient at Norristown to gather more information for his unpublished book.

230

With these words, he became the first gay psychiatrist to publicly discuss his sexual orientation. Fryer also revealed that there were many others like him, over a hundred, who belonged to the APA as psychiatrists. This shook up the self-protective, insular field. (Fryer, however, would not publicly reveal his identity as "Dr. Anonymous" for another twenty-two years.) Heterosexual psychiatrists could not imagine that one of their own could have such a debilitating "dysfunction."

On February 1, 1973, mere weeks after "On Being Sane in Insane Places" was published, the APA's board of trustees called an emergency meeting in Atlanta to address the many thorns in the side of the profession. Chief among them: the "deep concerns over rampant criticism that attend psychiatry today" (ahem, Rosenhan). The key outcome of this special policy meeting was to revise the *DSM-II*. Later in 1973, the APA sent questionnaires out to psychiatrists asking them whether or not homosexuality should be included in the *DSM* as a psychological disorder (you can't make this stuff up). Even to those who supported removing it, the idea that an "illness" could be stricken out with a survey showed how flimsy the whole operation was, and further supported

Rosenhan's theory that psychiatry's diagnostic system was arbitrary and unscientific.

Columbia psychiatrist Robert Spitzer, then a junior member of the APA's Committee on Nomenclature, joined the effort to redo the *DSM-II.* His first task was to define. "If you're going to have some people saying homosexuality is not a mental disorder, well, then what is a mental disorder?" Spitzer asked. He scoured the *DSM-II* to see if any tie bound all of the conditions. "I concluded that the solution was to argue that a mental disorder must be associated with either distress or general impairment," Spitzer later said. Around that same time, a secret group called the Gay Psychiatric Association invited Spitzer to sit in on a meeting, and this interaction provided the tipping point. If such successful people — without any obvious distress or impairment — could be gay, then how could they call it a disorder? The outcome of this revelation was that the APA scrubbed homosexuality from the new edition of the *DSM* — though traces remained in the diagnosis "Sexual Orientation Disturbance," which described people distressed by their sexuality (which, frankly, was probably anyone who was gay during a time when it was considered criminal and ill). A local newspaper satirized the removal with the headline: "Twenty Million Homosexuals Gain Instant Cures." Other interest groups took note: Veterans lobbied for the inclusion of post-traumatic stress disorder and got it in the manual in 1980; at the same time, feminists expressed their own concerns about diagnoses like "self-defeating personality disorder," a victim-blaming illness category, they argued, that provided scientific basis for patriarchal oppression. "Not only are women being punished (by being diagnosed) for acting out of line (not acting like women) and not only are traditional roles driving women crazy," wrote psychologist Marcie Kaplen, "but also male-centered assumptions — the sunglasses through which we view each other — are causing clinicians to see normal females as abnormal."

Psychiatry didn't even try to cover up its freak-out.

All around them, other scientists were colonizing space, transplanting hearts, giving deaf people the gift of hearing with cochlear implants. Physicians reported successfully transplanting bone marrow from one woman to another with Hodgkin's lymphoma. Mammography gave doctors a noninvasive way to look inside the body to detect breast cancer. We were mastering the great mysteries of the world — conquering space, cancer, and infertility. But we still couldn't properly answer this question: *What is a mental illness?* Or better yet, *What isn't?*

13
W. UNDERWOOD

This was an exciting time for those who demanded a revolution in psychiatry's ranks, and Rosenhan and his study stood on the front lines. Yet, strangely, at the height of his success, Rosenhan began pulling back from the spotlight. Why, for example, had he never finished his book? He had landed a lucrative book deal (the first paid installment, eleven thousand dollars, was the equivalent of an assistant professor's yearly salary) and had even written eight chapters, well over a hundred pages of it. By 1974, Rosenhan had already shared several chapters with Doubleday book editor Luther Nichols, who was enthusiastic and hungry for more details. In an editorial letter, Nichols promised that success was all but assured. "More work of this kind will get the book finished before you know it," Nichols wrote, "and then, if present interest can be sustained and certain features enhanced as described above, some very pleasant rewards should come your way. They will be well deserved." But Rosenhan would never reach out for these "pleasant rewards." He achieved what few academics ever do — worldwide attention and adoration, earning him a spot among the greats of the field — but in his son Jack's words, the study "became the bane of his existence."

This sudden instinct to shun the spotlight fell in line with other quirks from his private papers that I couldn't quite square. He took such pains to keep the details of the study a secret that he even used the pseudonyms in his personal notes. Who was he trying to protect?

I returned to Palo Alto and visited Rosenhan's son, Jack, hoping he might be able to lead me to some clues to better understand his father's motivations. Jack, the kind of teddy-bear man you can't help but hug the first time you meet him, adored his father, but freely admits that he doesn't share David's love of academics. Jack is an active guy with a contagious laugh; a man whose talents veered outside the classroom and onto the fields, more comfortable in tracksuits and baseball caps than suits and ties. Jack loves his family — his two girls; his wife, Sheri — and the soccer team he has coached to state championships.

We sat at his dining table as Jack spread out pictures, letters, and books from his garage — content that I had not yet seen — that had survived his father's move to a nursing home over a decade ago. Jack shared stories of his father's sharp humor and his gentle but firm parenting style. Jack recalled the time he sneaked out to go to a party when he was a teenager and returned to find every entrance to the house locked except the sliding door to his parents' bedroom. When he stepped in he found his father wide awake, greeting Jack from his bed, asking if he had a good time and to please close the sliding door behind him. Jack lay awake all night worried about the trouble he had gotten himself in, but the next morning his father wasn't mad — in fact, Rosenhan gave him a later curfew. Jack was so unnerved by the experience he never sneaked out again.

We sifted through Jack's photo albums: a picture of Rosenhan with Jack at his wedding, their arms outstretched in a gesture of celebration, Rosenhan's beard flecked with gray and Jack young and rosy-cheeked; Rosenhan during his graduation from Yeshiva University wearing a cap and gown, black-rimmed glasses, and a mischievous grin; Rosenhan in his twenties goofing off for the camera; Rosenhan and Mollie on their wedding day; Rosenhan as a child smiling broadly with his scowling, buttoned-up mother and his equally smiley younger brother. A life.

While sorting through the boxes in his garage, Jack discovered a few more diary entries from Rosenhan's Haverford hospitalization and letters from his stay that were addressed to Jack. On cursory glance, the letters looked like David's other handwritten notes — beautiful but barely legible, and coded.

And then a clue.

I nearly discounted it, thinking it was yet another outline of his unpublished book *Odyssey into Lunacy,* until I saw that this version was handwritten, unlike the typed versions in his files. Next to a bullet point reminding him to add references to a study, Rosenhan had written: "see list [?] sexual preoccupation (I owe this to W. Underwood)."

W. Underwood. The name sounded familiar, but I had cycled through so many names over the course of my research that it was impossible to pin down the source. It wasn't until weeks later, scouring my files, that I

238

came across a list of psychology graduate students photocopied from Stanford's 1973 yearbook during an earlier visit to the campus's Green Library. And there was W. Underwood.

A PubMed search for "Wilburn Underwood" yielded a clear link to David Rosenhan. In 1973 and 1974, a Wilburn Underwood and David Rosenhan co-authored two studies on affect and altruism in children, measuring how charitable second and third graders would be when primed to be happy or sad by rigging a game so that each child was a "winner" or a "loser," the same bowling game Rosenhan had used in his research on children at Swarthmore. The second-listed author, a man named Bert Moore, gave me a clear-cut lead: He worked as the dean for the School of Behavioral and Brain Sciences at the University of Dallas. I shot off a quick plea for help, realizing that it was a long shot that Bert would remember a man he'd worked with four decades ago, let alone still be in touch with him.

To my delight, Bert returned the email within minutes with contact details for "Bill." I would later learn that Bert Moore sent me this email while suffering through the final stages of pancreatic cancer.

Bill — I now had a first name, which

matched Rosenhan's description of the soft-spoken, red-bearded graduate student named "Bill Dixon." Dixon was, according to Rosenhan, "the person least likely to make it through the admission interview. Professors should not be trusted to objectively evaluate their students. But for what it is worth, Bill struck me then as he does now, as a person with an enormous sense of balance. He works very hard, and he plays equally hard." There wasn't much else written about Bill, but this whisper of the man seemed pretty compelling to me.

I tamped down my growing enthusiasm, reminding myself that Bert hadn't confirmed that Bill Underwood was a pseudopatient — just that he existed and was still living somewhere with a Texas area code. I wrote to Bill and five days later, on my birthday, I received this gift:

Hi, Susannah.
I was indeed in the pseudo-patient study. I can't imagine what I would have to add but if you want to talk that would be fine.

Bill U

There he was. My first living pseudopatient.

14
CRAZY EIGHTS

A month later, I rented a car at Austin-Bergstrom Airport and headed off to the Underwoods' home in the Austin Hills. I rolled the windows down to take in the oppressive Texas heat, a relief from the East Coast's never-ending March frigidness, and tapped my foot to the sounds of Tom Petty as I turned into the Underwoods' driveway.

I steadied myself outside the house, overwhelmed with the collywobbles, a feeling that I recognized from earlier days working as a news reporter for the *New York Post.* I still get nervous before interviewing strangers, but I know enough now to recognize those nerves as a good sign. Without them, I'll fumble.

Bill Underwood and his wife, Maryon, invited me in, offered me tea, and pointed me to their comfy white couch. Bill summarized his career after Stanford. He graduated the same year that the study came out, took a position at Boston College as an assistant professor, and then moved to Austin to work at the University of Texas as a psychology professor. When he didn't receive tenure he returned to school, this time for engineering. He landed a job at Motorola as part of their research team and had recently retired from a software company. In that time he had folded the study away, his contributions to the history of psychology destined to remain unknown.

Wilburn "Bill" Crockett Underwood was born in West Texas on July 30, 1944, while his father was stationed at a naval base in Hawaii in the aftershock of Pearl Harbor. His unusual middle name came from his father, who went by Crockett, a nod to the family lore of a distant kinship with the king of the wild frontier, Davy Crockett. When his father went off active duty, the family moved to a small, oil-rich town on the Gulf Coast called Mont Belvieu, Texas, made up of mostly blue-collar oilfield workers, rice farmers, fishermen, and, most important, Bill's high school sweetheart and future wife, Maryon. Bill graduated as valedictorian, which, he said in his laconic manner, "really wasn't that hard to do," competing against only eighteen other kids. After high school, the couple left the small town and never looked back. Bill enrolled at the University of Texas at Austin, where he received a degree in mathematics but developed an interest in psychology. Maryon, meanwhile, gave birth to the first of their three children.

To make extra cash, Bill worked the graveyard shift as an attendant at Austin State Hospital (much as Ken Kesey did during the writing of *One Flew Over the Cuckoo's Nest*). Bill's shift started at 11 PM, so most of the patients were asleep by the time he arrived and were just waking up when his shift ended. He killed time arranging medications in little paper cups so that the nurses could easily dispense them the following morning. His nights, though "interminably boring," allowed him to peer into the gradations of madness — from alcoholism to full-blown psychosis. One man in particular, who refused to walk anywhere near windows because he believed that airplanes were taking pictures of him, made a particularly strong impression on Bill. These delusions were *real* to him, as real as the words on this page are to you. After three months, Bill gave up his shift when the night hours weighed too heavily on him and his growing family.

During the day, Bill and Maryon attended

classes at the University of Texas, Austin. Maryon was on campus that fateful mid-morning on August 1, 1966, when Charles Whitman climbed the tower with his hunting rifle. She remembers the details as if they happened yesterday. Pretty Maryon must have made quite the spectacle in her neon-yellow wraparound mini-skirt as she walked across campus to the parking lot, released from class a few minutes early. When she arrived at student housing, she heard frantic rumors about a gunman. Some people had heard that the shooter was on top of the tower, others that he was traveling from building to building. There were no protocols because this had never happened before. People didn't know whether they should hide or flee.

Earlier that morning, Whitman, a twenty-five-year-old ex–Marine Corps engineering student, killed his mother and his wife, then filled a footlocker with rifles, a sawed-off shotgun, and handguns, stopped at a local gun store to buy boxes of ammunition, and headed to the UT Tower. He took an elevator to the top and climbed the stairs to the observation deck, shooting three people at point-blank range. He then set up his arsenal and aimed his sights on a pregnant woman. Next, her boyfriend walking with her.

Whitman left behind a suicide note. "I don't really understand myself these days," he wrote. "I am supposed to be an average reasonable and intelligent young man. However, lately, I can't recall when it started, I have been a victim of many unusual and irrational thoughts . . . After my death I wish that an autopsy would be performed on me to see if there is any visible physical disorder."

Whitman murdered seventeen people. Eventually, two Austin police officers intervened, shooting Whitman dead. An autopsy revealed a glioblastoma, a malignant tumor the size of a nickel that was growing beneath his thalamus and against the amygdala, associated with fight-or-flight responses and highly implicated in our expressions of fear and anger.[1] Though it's unclear if this

---

1. The same thing still happens today. When Stephen Paddock committed suicide after opening fire on a concert in Las Vegas in 2017, killing fifty-eight people and wounding five hundred, authorities shipped his brain to Stanford in an effort to track down any biological basis for such unthinkable evil. As of this writing, Stanford has not released the results.

caused him to snap and terrorize a campus, there was a "palpable sense of relief" when that tumor was discovered, Bill recalled.

"We all wanted there to be a reason for him to have done what he did," Maryon added. If there was something biological — in other words, something that could explain why — it would soothe many souls. Simultaneously, though, it raised the inevitable question: Could we all be just a tumor away from shooting up a college? Maryon remembered waking up in the middle of the night and looking at her husband. "For that moment before I could calm myself down, I was terrified of him. I mean, how well do we know anyone?"

Charles Whitman's story underscores, yet again, the ever-present appeal of finding objective measures that can separate illness from wellness. Soon after Whitman's rampage, new technologies promised easier and more sophisticated access to the brain. Imaging took off in the early 1970s, starting with the invention of CT scanning, allowing us for the first time to peer inside our living skulls. Older techniques were crude and dangerous, and involved draining the cerebrospinal fluid via a lumbar puncture and replacing the fluid with air, a technique used only in the direst situations. Now researchers and clinicians could scan anyone. A flurry of brain studies followed, leading to advancements in the understanding of the palpable differences between "sick" and "healthy" brains at the level of structure — such as enlarged ventricles (the cavities in the brain where cerebrospinal fluid is produced), gray matter thinning in the frontal lobes, and volume reduction in the hippocampus, sometimes seen in those with serious mental illnesses, like schizophrenia. All of this coincided with the research revolution in neurochemistry and contributed to the supremacy of the biological model of mental illness.

But the hope that CT scans would provide a laboratory test to diagnose schizophrenia crash-landed as follow-up studies revealed that many people diagnosed with schizophrenia did not have, say, enlarged ventricles compared with healthy controls, and that some people with bipolar disorder and "normal" controls *did* — which undermined the diagnostic significance of these findings. More advanced imaging technologies emerged, like PET scans and MRI, promising, as neuroscientist and psychiatrist Nancy Andreasen wrote in her optimistic 1984 book *The Broken Brain,* that the biological revolution in psychiatry would solve the "riddle of schizophrenia . . . within our lifetime, perhaps even within the next ten to twenty years." We're still waiting.

Everything from sustained antipsychotic use to smoking cigarettes to childhood trauma changes the brain, making it hard to disentangle exactly where the disorder begins and environmental factors end. In 2008, researchers for the journal *Schizophrenia Research* conducted a literature review of all the relevant articles on schizophrenia published between 1998 and 2007 — over thirty thousand of them — and found that "despite vigorous study over the past century . . . its etiology and pathophysiology remain relatively obscure and available treatments are only moderately effective." Little has changed in the ten years since. This isn't surprising given that the brain is a protected organ, isolated from the rest of the body and nearly impossible to study in real time.

The brain didn't interest Bill, however, as much as the social behavior research by Stanford professor Walter Mischel, the author of *Personality and Assessment.* So he applied to Stanford to work with Mischel. Bill's daughter Robyn even participated in Mischel's marshmallow tests on delayed gratification, the series of studies that made Mischel a (near) household name. For it, researchers gave three- to five-year-old children from Stanford University campus's Bing Nursery a treat, a marshmallow in most cases, and told them that if they could wait a few minutes without eating it, they would be given a second one. Mischel found that a child's ability to show restraint in the face of a fluffy treat correlated with later measures of IQ, higher SAT scores, lower body fat percentage, fewer behavioral issues, and greater sense of self-worth. (All Robyn remembers is sitting at a table with peanuts and mini-marshmallows. She doesn't remember if she was able to delay her sweet-tooth urge or not.)

Stanford wasn't exactly Berkeley, but it was still California in the late 1960s, and somehow the Underwoods settled into the chaos. They joined protests, staffing phones and distributing leaflets for an organization called Movement for a New Congress, and helped peaceably intervene in a battle between rock-throwing protesters and the National Guard. Bill tooled around on his Yamaha two-stroke motorcycle and listened to Jimmy Cliff records. The Underwoods don't like to admit it today, but they were cool.

In the fall of 1970, Bill signed up for Rosenhan's seminar on psychopathology. Bill adored Rosenhan from moment one, using words like "charming" and "charismatic" to describe him. "When you talked to David, you felt like you were the most important person in the world," Bill said. Small seminar classes showcased Rosenhan at his most riveting, especially when he lectured about his time undercover as a patient. It was only in the retelling that Bill realized Rosenhan was recruiting. He was subtle about it, but his intention was clear, at least in retrospect: "You would want to be involved in almost anything that David was doing," Bill said.

I was a little surprised, I admit, by Bill's characterization of how little preparation went into his hospitalization, which was not the way Rosenhan portrayed the process. Rosenhan talked about weeks of prepping: going over backstories, teaching data collection methods, establishing the basics of life on the ward, but Bill recalled none of this. Rosenhan showed him how to cheek pills, which was basically: "You just put it in your mouth, close your mouth, slip it under your tongue, sip the water, walk aimlessly around for a couple of minutes, and then go into the bathroom and spit it into the toilet,"

Bill said. It wasn't exactly thorough advice; nor was it airtight.

Perhaps this was why Craig Haney, then a teaching assistant in Rosenhan's psychopathology class who later worked with Philip Zimbardo on the famous prison study, declined Rosenhan's offer to pose as a pseudopatient. "I didn't want David to be my lifeline," he said. But Bill saw it all through rosy, Rosenhan-filtered glasses. "The idea was that you go in and sort of experience it cold turkey as it were."

Bill came up with the last name Dickson, a subtle dig at President Nixon (which explains why Rosenhan had misspelled Bill's pseudonym as Bill Dixon in his notes, adding another layer of misdirection to my search for the others), and established a backstory. Bill remained a student but dropped his psychology focus and also his marriage so that if things went awry, a distance remained between the real Bill and the fake one.

Like Rosenhan, Bill didn't actually believe he'd be admitted. In his book, Rosenhan repeatedly emphasized that Bill was "least likely" to be admitted because he was "a person with an enormous sense of balance." His good humor, dry wit, and placid demeanor — his utter *solidness* — made it

252

seem impossible that any psychiatrist would commit him. Maryon wasn't as confident. "I was a nervous wreck," she told me. Her imagination ran wild with images from the movie *Snake Pit,* where patients were neglected, shocked, and abused.

Bill had conducted enough research to know that Agnews State, which Rosenhan called Alma State, the hospital Rosenhan had chosen for him, didn't just take in people off the street. He first had to drive twenty minutes to a community mental health facility in San Jose, where he would be observed to see if hospitalization was necessary, a new layer of protection added by the Lanterman-Petris-Short Act that was signed into law in 1967 by then-governor Ronald Reagan. The act, which went into full effect in California in 1972, intended to make it much more difficult to involuntarily hospitalize patients or hold them for an extended period of time.

Bill made no effort to "look the part" — he wore a clean T-shirt and bell-bottoms. The bushy beard remained, as did his longer, slightly wavy hair and thick black-framed glasses. The interview went as planned: Bill told the intake officer that he was a student at Stanford, that he was unmarried, and that he had started hearing voices, sticking strictly to the script, saying that he heard them say "thud, empty, hollow." His nerves probably helped sell his story. The interviewer handed over his case file and told him to find a ride to Agnews State Hospital, where he would be admitted.

Bill asked Maryon to drop him off out of the sightline of Agnews State Hospital's front entrance for fear that . . . what? That someone would see him with a woman and assume he was lying about having a wife? (This seems pretty paranoid to me. I think that the shock of his admission hit him harder than he admits.) Maryon watched as her husband walked up the palm-tree-lined drive to the entranceway of the stately psychiatric hospital. Right then, she said, she knew he wasn't coming back.

Bill's dread deepened the closer he got to the admissions building. Eventually he reached a sign directing him to INTAKE, which looked like an ordinary doctors' office waiting room, where patients were diagnosed and sent off to hospital wards that had become increasingly ill equipped to deal with them.

Located less than half an hour south of Palo Alto in the city of Santa Clara, the Great Asylum for the Insane (later renamed

Agnews State Hospital) opened in 1885 after a farmer donated his three-hundred-acre farm to the state to house the growing army of the "chronically insane." Superintendent Leonard Stocking, who lived on the grounds, instituted a return to a more humane approach to psychiatric care called moral treatment (which, as we saw earlier, proliferated in the 1800s until it over-reached). Stocking built libraries, gymnasiums, a piggery, and a chicken coop, and opened tracts of farmland, all maintained by patients and staff. His daughter Helen Stocking lived on one of the wards for most of her adult life and even wrote and directed plays that patients staged in her honor.

But Agnews, like most institutions, was a product of its time, and the institution where Maryon dropped her husband off was not the same place Helen Stocking lived and wrote. "They were tense times," former Agnews psychiatrist Izzy Talesnick told me. Money was tight, and the hospital was plagued by the lethal combination of over-crowding — at its height it held forty-five hundred patients — and understaffing.

Upon arrival, Bill participated in a series of interviews. A German Nurse Ratched type interrogated him about his sexual preferences and drug use. Rosenhan quoted from Bill's notes, which Bill told me he threw away years ago. "A woman who had only limited command of English talked at length about my sex life. She pressed me for an admission of homosexual activity. She also asked about my childhood more than the others did. She asked if I had been jealous of my father."

Bill's beard, his long hair, and his clothes created a portrait of the perceived "other," a mentally ill deviant, which at that time was a gay man. He continued: "They seem to want to press me into admitting the use of psychedelics." This was yet another example of a doctor seeing what she expected to see. We witnessed it with Rosenhan, when the doctors described his "constricted speech." This type of misjudgment is common in physicians; it predisposes people to fill in the unknowns and disregard anything that may not support their conclusions.

It took the admitting psychiatrist less than half an hour to reach her diagnosis: paranoid schizophrenia. He was officially admitted — case #115733.

Bill was placed into a dorm with twenty other men. He was now just one grain of sand in a desert of sick men, as if he had always been there and always would be. The unwritten rule was that you never asked "Why are you in?" Diagnoses were rarely, if ever, discussed, though everyone knew the difference between the "acutes," or the temporary ones, and the "chronics," who were lifers. There were guys in for drugs and alcohol, those who went on a few too many acid trips or — more frighteningly — did one acid trip and lost it; there were some McMurphys there, too, malingerers who were there to dodge the draft or escape from their lives. Sometimes Bill mistook the staff for the patients, until he noticed their keys, a signal of distinction that Rosenhan also noted, which separated "them" from "us."

Bill made a friend whom he nicknamed "Samson." All Samson talked about was his hair. He felt his power and mental strength were forged in his follicles. Sure, hair was important. Bill had grown out his wavy red hair long enough to put up into a ponytail to announce where and how he fit into the new world. But this was something altogether different. Samson had started dealing drugs, and to make his new career a little less obvious to narcs, he had cut his hair off. When the drug deal fell through, and Samson realized that he had chopped it off for nothing, he attempted suicide. He survived and ended up on Bill's ward. Magic hair aside, Samson made sense to Bill. He was the kind of guy you might see around campus. The two spent hours talking and playing cards — of all games, crazy eights.

In her husband's absence, Maryon's mind drifted off into dark places. She couldn't bat one particular image away: men strung up by their ankles from the ceiling. Where she got this, she doesn't know to this day. She tried to focus on her girls but lost herself in crying jags. *Will they medicate him? Shock him? Tie him up in a straitjacket?* Her friends and neighbors acknowledged her red eyes and Bill's sudden disappearance but didn't pry, assuming that the couple had hit a rough patch. All she could do was brush them off. She had promised Bill and Rosenhan not to tell a soul.

A day later, on Friday the thirteenth, she was finally able to visit. Walking up the same palm-tree-lined walkway that she had watched her husband disappear along, she felt almost outside herself as she asked the receptionist for "Bill Dickson."

Door. Hallway. Door. Second hallway. Door. A huge, double-wide oak door the size of something you might find on a college campus.

She heard scratching on the other side. She pictured patients clawing the door, their fingers bloody nubs where their nails should be, desperate to be freed. As the door swung open, she recoiled, bracing herself for the worst of her visions.

But there was only David Rosenhan. The scratching sound came from Rosenhan fiddling with the locks (somehow he had a key).

"How is he?" Maryon blurted out. Rosenhan was her one source of calm. He had been so kind to her in her husband's absence, advising her to write down her thoughts in a journal since writing had helped him during his own hospitalization. He reassured her that Bill was safe thanks to writs of habeas corpus that he had filed. The idea that a piece of paper was prepped and ready to go that could release her husband soothed her.

I interrupted Maryon here. The writ of habeas corpus — the term is Latin for "that you have the body" — is the document that saved Elizabeth Packard from false imprisonment in the 1800s. Once presented, it required that Bill be brought before a court, where it would be determined if his hospitalization was valid. Though Rosenhan did write in "On Being Sane in Insane Places" that "a writ of habeas corpus was prepared for each of the entering pseudopatients and an attorney was kept on 'on call' during every hospitalization," this wasn't entirely true. I had tracked down the ACLU lawyer named Robert Bartels, now based in Arizona, who had worked as a law assistant aiding Stanford professor John Kaplan with Rosenhan's experiment. Bartels was a bit hazy on details, but he was confident that though they had discussed writs for one or two people, he had never prepared any and that "on call" may have been an exaggeration. When I told Maryon this, her anger flared. "Good thing that I didn't know — that's what got me through. I guess that I was naive. I just believed."

Back in that doorway: She didn't remember what Rosenhan said, only that he looked distressed. And then he was gone. Maryon found herself on the other side of the locked door she so dreaded. Did Rosenhan tell her how to get there? She doesn't recall. The next minute she found herself in the dining room, which reminded her of her high school cafeteria, her thoughts resting on a safe place, on Bill, her high school sweetheart.

There he was. Bill was slumped down in his seat, his head resting on his folded arms.

He seemed to be either crying or dead asleep. She approached the table and softly called his name. He didn't budge, didn't even acknowledge her presence. She took the seat opposite her husband. Eventually he lifted his head. "I'm sleeeeeepyyyyy," Bill said. His words came out muddy, as if he'd had a few scotches too many. Forget the hanging bodies or the bloody fingernails. *This* was the real fear. Her husband was *altered.*

An hour or so before Maryon's visit, a nurse clad in stiff whites had walked through the cafeteria handing out paper cups with pills. When she handed one to Bill, he recognized the medication from Austin State: Thorazine, psychiatry's miracle drug. Bill had felt confident that he could easily cheek the pill. He popped one without thinking and let it nestle beneath his tongue. But what he didn't expect was the burning sensation. The new capsule coating was designed to melt away, making him feel like it would burn a hole in his mouth if he didn't swallow. He stumbled toward the nearest bathroom, but didn't make it there before his automatic reflex took hold and he swallowed. Bill was well aware of the drug's side effects — the tremors, nonstop drooling, uncontrollable body movements, muscular rigidity, shuffling gait, and blue tinge of the overdosed — and he comforted himself with research he had read in class about the placebo effect. He had to *believe* that all would be okay for it to be. But when he finished his meal and walked out into the ward, the world went black.

Next thing he knew he was being shaken awake by an attendant, who told him it wasn't time to sleep. He had a visitor. David Rosenhan.

Bill told me he didn't remember their conversation, and Rosenhan didn't write about it. Rosenhan kept spare notes on Bill's hospitalization, which were mostly found in a few short sections of his unpublished book. All Bill could recall was an unrelenting desire for sleep. "I would have paid a thousand dollars right then and there to just put my head down," he said.

"Did he notice that you were . . . did you tell him that you had taken the drug by mistake?" I asked.

"I don't think I did."

"Did he notice that something was off?"

"I don't know. He didn't say. He didn't say anything if he did. I may have made more of an effort to hide that with him than I did with Maryon. That's one of the nice things about being in a relationship, you don't have to hide that stuff."

This was why Maryon had found him so changed. "I was used to being married to somebody that was going to have a PhD someday," she told me. "Somebody who had control over his life, had control of everything. To see him in a situation where he was like an invalid almost, where he couldn't do anything or make decisions, that was hard."

This institution had suddenly transformed her husband, and she didn't know when — or if — she would get him back.

15
WARD 11

As Bill shuffled the deck for another round of crazy eights, a miraculous series of events was unfolding a few yards away inside the same hospital on a special unit called Ward 11.

The idea for Ward 11 was sparked in the mountains of Big Sur at Esalen Institute. Most people of a certain age know of Esalen thanks to its notoriety — Naked therapy! Orgies! Drugs! (And, more recently, many may recognize it as the setting of *Mad Men*'s finale episode, where Don Draper experiences his "I'd like to buy the world a Coke" aha moment.) Two years before Bill's hospitalization, a *Life* magazine article skewered Esalen. It reads like satire: "Not only do people publicly neck and nuzzle like teenagers, but they sit on each other's laps like babies. And they cry a lot. Crying is a sort of status symbol."

Despite the bad press, Esalen was a key

incubator for the growing counterculture and human potential movements as everyone from movie stars, businessmen, and bored housewives tapped into their better selves. Attendees participated in programs like "The Value of the Psychotic Experience." Bob Dylan visited. R. D. Laing lectured. Joan Baez was basically an artist in residence. Charles Manson showed up with one of his girls and performed an impromptu concert days before the Tate murders. During the heady first decade, you may have rubbed shoulders with anyone from British philosopher and Eastern culture disseminator Alan Watts; to chemist Linus Pauling, one of the founders of quantum mechanics and molecular biology; to writer Ken Kesey; psychologist B. F. Skinner; and quite possibly to social psychologist David Rosenhan. Despite the debauchery and celebrity worship, the goal — to offer a peaceful oasis away from the world's soul-crushing conformity — was a legitimate one dreamed up by Esalen cofounders Mike Murphy and Dick Price, who had barely survived his experiences on the other side of sanity.

Dick Price was supposed to follow in his successful father's footsteps: attend a respectable school, major in economics, and settle down with a suitable wife. Instead he pursued a degree in psychology and developed an interest in Eastern religions after taking a class by Frederic Spiegelberg on the Hindu text Bhagavad Gita, which championed the pursuit of a "dharma" or path that each enlightened person is destined to fulfill. He seemed back on the straight and narrow when he enlisted in the air force — if you ignored the fact that he spent his nights at The Place, a nightclub in San Francisco's North Beach neighborhood, frequently haunted by Allen Ginsberg and poet Gary Snyder. Soon Price met a dancer and fell hard. On the night he met her, he heard a disembodied voice say: "This is your wife." The two married. It all sounded poetic, even though it was the beginning of Dick's unraveling.

His behavior grew odder, even within the affectedly strange and drugged-out Beatnik scene. One night at a bar in North Beach he was hit by an urge: "He felt a tremendous opening up inside himself, like a glorious dawn," wrote political scientist and author Walter Truett Anderson in his book *The Upstart Spring*. The feeling was: "I'm a newborn, I should be celebrated." Price began repeating: "Light the fire, light the fire," over and over, spooking the bartender, who called the cops. Price ended up in handcuffs and woke up in a psychiatric hospital at the Parks Air Force Base, where he fought aides and was sequestered in a padded isolation room. He threw himself against the walls, believing that there was an "energy field" around him that protected him from injury and pain. There he received his first in a series of electroshock therapies.

Dick's family moved him to a fancier private hospital across the country in Hartford, Connecticut, called the Institute of Living. On the surface, the institute had more in common with a country club than a hospital. A Victorian main mansion, surrounded by cottages and research buildings, stood on ornate grounds designed by Frederick Law Olmsted, the chief architect of Manhattan's Central Park. Patients could pick from a fleet of chauffeur-driven Packards, Lincolns, and Cadillacs. There was even an in-house magazine, *The Chatterbox,* which once ran an illustration of glamorous patients wading around the pool.

But these images only told the stories the institute wanted to share. Though the hospital catered to the rich and famous with its putting greens and fancy cars, it also deployed the experimental treatments of the era — lobotomies, ECT, and insulin coma therapy. The institute's psychiatrist-in-chief, Dr. Francis J. Braceland, had deep attachments to the Catholic Church and admitted priests who had been sent by archdioceses to be "cured" of their "disorders." Pope Pius XII knighted him in 1956, the same year that Dick entered the hospital, where doctors diagnosed him with paranoid schizophrenia.

At the Institute of Living, Dick lived on the locked ward, his "private prison," where he was subjected to cutting-edge "treatments." During his stay he underwent ten electroshock therapies, doses of Thorazine, and what Dick called "the complete debilitator," insulin coma therapy. Viewed at best as malpractice, this therapy, which involved inducing comas with insulin to cure psychosis, went out of style by 1960 after a series of articles revealed that there was no scientific evidence to back the dangerous, sometimes lethal procedure.

This is what Dick would have faced: After a series of tests — blood work, heart rate monitoring — a nurse would inject the insulin. As his glucose levels fell, Dick would sweat and salivate; his breathing would slow and his pulse quicken. Gradually, unconsciousness would blanket him. Patients would sometimes drool so much that nurses would have to sop up the saliva with sponges. Sometimes the skin burned hot, the muscles twitched, and the patient would jerk. Often a seizure would occur, which doctors saw then as a sign that the treatment was working. Glucose injections followed, administered intravenously or via a thin rubber tube inserted through the nose and into the stomach, bringing the patient back to life (if they were lucky).

During the year he spent at the institute, Dick Price said he underwent fifty-nine of these therapies. The naturally trim Price, a born athlete, put on over seventy pounds, since the insulin treatments caused ravenous hunger. He fell into a stupor, wandering around the halls as if he were in "a pool of molasses" until something clicked inside him: He had to get out. After learning how to successfully cheek his Thorazine, Dick convinced his father to get him off the locked ward and onto an open one. On Thanksgiving Day 1957, he was released. (Another famous Institute of Living resident, screen actress Gene Tierney, would later call her stay there "the most degrading time of my life . . . I felt like a lab rat.")

Dick Price returned to California, where he hooked up with Mike Murphy, whose family owned the land on which the two

men would build Esalen, their dream re-
treat, which they opened to the public in
1962. Price envisioned Esalen as a place
that "would serve people coming to this
type of experience and there would not be
the drugging or the shocking — that was
my main motivation." He believed that
madness should be taken seriously, probed,
embraced, and examined as a path to insight. He saw Esalen as a place to "live
through experience" and facilitated this approach by providing treatments like encounter therapy, bodywork (massage, Rolfing,
and sensory awareness), and psychedelic
drugs. Dick was influenced by the work of
Fritz Perls, a German psychotherapist in
residence at Esalen, who created Gestalt
therapy, which pushes people to focus on
the present moment.

R. D. Laing came to Esalen in 1967,
speaking in his enchanting Scottish brogue
about his work at Kingsley Hall, a house in
the East End of London that provided
therapeutic supportive housing as an alternative to hospitalization. Kingsley Hall was
a utopian place, Laing said, with no de-
individualization, no power struggles over
keys, no forced meds, where people engaged
in twenty-four-hour therapy sessions and
meditated. (He didn't mention the young

woman who smeared feces on the walls, the LSD sessions, the drug raids, or the parade of celebrities who gawked at the scene, but that's a different story.)

That same year psychologist Julian Silverman, a National Institute of Mental Health schizophrenia researcher, arrived at Esalen to teach a seminar on "Shamanism, Psychedelics, and the Schizophrenias." He wasn't your typical buttoned-up doctor type. Silverman had befriended the Grateful Dead and followed the teachings of John Rosen, the inventor of "direct analysis," which used psychotherapy to treat schizophrenia by basically babbling with the patient. (Rosen later lost his license after patients accused him of sexual and physical abuse, landing him on the long list of doctors who exploited people under their care then and now.) Silverman and Price hit it off, and out of their friendship grew Ward 11, a way to scientifically test Laing's therapeutic housing theories.

Dick Price offered to supply funding from Esalen's coffers, and the National Institute of Mental Health supplied grants. Somehow they convinced Agnews to allow them access to a ward where they could conduct the experiment. Maurice Rappaport and Voyce Hendrix (yes, he's a close relative of *that* Hendrix) joined in to work in "ding dong city," as Silverman lovingly called it.

They selected a few Agnews staff members, vetted to be young, far out, and open-minded, to travel to Esalen to learn Gestalt therapy. The staff discouraged separation created by "the cage" and set up a quiet vigil room, where anyone who felt overwhelmed could get away and sit and pray or just think. Staff members were to interact with the patients as much as possible. Patients were allowed to roam freely throughout the ward — a big no-no in most hospitals, which steer patients into dayrooms so that they can be watched from the cage. The criteria were simple: Men between the ages of sixteen and forty, recently diagnosed with schizophrenia without prolonged history of mental illness, would live on Ward 11. They wanted patients with no prior hospitalizations — most of them were "first breaks." Half would receive nine tablets a day of the typical course of Thorazine, a minimum of three hundred milligrams a day, while the other half would receive a placebo. (Interestingly, Bill Dickson himself, who was in a nearby ward, did fit the above criteria. It's possible that he was considered for inclusion in the study, but he is confident he was never included.)

The beginning was rough, to put it mildly. "The first thing we did was take some patients from the hospital and take them off medication. They broke all the windows the third day," Alma Menn, a social worker on Ward 11, told me.

The new freedom created some friction, it seemed.

"We only really had one fire," Alma added.

The fire occurred during a visit by a psychotherapist, who lugged in a bin of toys, dolls, and musical instruments to facilitate playacting with adults. The staff rifled through the props alongside the patients. That's when the fire department walked in.

"Of course I had been holding my skirt on my head and playing like a mermaid. We all had an instrument and were playing music," Alma said. "[The firemen] walked around the corner and there was a patient who'd been in bed and was standing at the drinking fountain with a cup trying to put out the fire that he had started on his mattress."

The result of all this playing was published in the 1978 paper "Are There Schizophrenics for Whom Drugs May Be Unnecessary or Contraindicated?" The paper showed that of the eighty patients studied, the placebo group showed greater improvement than its drugged counterpart, though both groups showed improved long-term outcomes over patients undergoing "typical" hospitalizations.

Rappaport's study added to the growing backlash against the "take your drugs" approach endemic to the traditional hospital settings. Patient groups, who now called themselves psychiatric survivors, had already started pushing against this refrain by filing class-action lawsuits against Big Pharma as many patients experienced permanent, disfiguring side effects. Suddenly these miracle drugs didn't seem so miraculous — in some cases they were downright dangerous.

Rappaport et al. gave an alternative approach scientific basis — though mainstream psychiatry successfully dismissed the findings as fringe science, missing the larger picture that the creation of a supportive environment had actually improved clinical outcomes for everyone. Something as simple as sitting and eating together, as listening, as goofing around, as playing dress-up, as being part of a community seemed to help.

Even though mainstream psychiatry ignored it, a series of "med-free sanctuaries" sprouted up around California. The most prominent person to take on the mantle of

Ward 11 was Loren Mosher, the head of the NIMH's Center for Studies of Schizophrenia, who saw an opportunity to take Ward 11 to the next level. He recruited Ward 11's cast of characters — including Alma Menn and Voyce Hendrix — to start Soteria House, an experiment in communal living located in a twelve-room Victorian house in downtown San Jose. Here a group of six people who would have ended up in an asylum lived together outside of it. The average stay was forty-two days — much shorter than the six-month average in an institutional setting — while the total doses of antipsychotic medications were three to five times lower. Papers published extolled the value of the environment and the success rate of using minimal antipsychotic drugs. As at Laing's Kingsley Hall, there was neither commitment nor forced medication. One of the board members who helped mold Soteria House — and here it all comes full circle — was David Rosenhan, in the midst of the success of his ground-breaking study whose theories questioned the powers of traditional psychiatry and its hospitals.

Over twelve years the outcomes of people living at Soteria, named after the Greek goddess of safety and salvation, varied. There were a few suicides. Some got worse and had to be hospitalized, but many reported that Soteria House was a transformative and ultimately healing experience. One former Soteria resident I interviewed credits his current life — he is a successful technology salesman with a wife and two children — to Soteria. It's easy to dismiss Soteria House, as many do (and as I did at first), but its mission captured something essential missing from the institutional model: focusing on the patient, not the illness.

The Soteria model continues in places like Alaska, Sweden, Finland, and Germany. There are echoes of it in the clubhouse model, which predates Soteria but provides similar restorative support, along with housing and employment opportunities to people living with serious mental illness. We see it also in Geel, a small town in Belgium with a long history of providing a safe haven to those with mental illness, where foster families in the community adopt "guests," not patients. In Trieste, Italy (where a young Sigmund Freud first studied the sexual organs of eels), people are respected as members of the community with access to care across a wide spectrum of needs, along with supportive social networks.

The legacy of Agnews's Ward 11 is a long

one. Sadly, Esalen's Dick Price probably did not get to bask in the immediate celebrations surrounding the launch of his successful research study. Just before the project commenced in 1969, Price suffered another break. He began ranting about "having more kingdoms to conquer," believing that he had channeled a whole host of historical figures, including Napoleon and Alexander the Great, and spent ten days in, of all places, Agnews State Hospital. Price did eventually recover and returned to Esalen, where he lived peacefully until his death in 1985.

16
SOUL ON ICE

Meanwhile, in the same facility, Bill's time on the acute ward was coming to an end. He spent forty-eight hours there before the hospital deemed him well — or rather, still unwell enough to be moved to the residential floor. The residential floor was less like a hospital, with lounge chairs and windows that lined the dayroom, giving it a "homier" feeling than the dark, gloomy acute floor. An outdoor space was open to those with grounds privileges (until a patient successfully jumped the wooden fence around it). Psychiatrists rarely visited the wards, and when they did their interactions with patients were swift and dismissive. A blunt male psychiatrist whose pointed questions bordered on the absurd had already been primed to ask Bill about his drug use and sexual orientation — foregone conclusions made by the previous psychiatrist, who had spent a mere half hour with him. Bill still received three daily doses of antipsychotic medication, but after that first incident in the cafeteria, he had learned how to properly dispose of the pills.

The other patients were like him, young hippies. Well, most of them. There was "the crawler," a young man in his mid-twenties who spent the majority of the day on his hands and knees navigating the grounds like a baby. "He was a very weird dude, obviously," Bill said. "But I was talking with some other guys at one point, and we were just standing around talking and he's crawling around. He crawls over to our area, gets up, and we start talking about college. He knew I was a college student, and he had been at junior college, community college somewhere in the area, and so we started talking about college courses, you know, and how hard it was and all that kind of stuff, and then we finished our conversation and he got back on his hands and knees and crawled off."

"Wow. It's kind of comical," I said.

"It is, but it's also . . . I mean, I think for a lot of people who are labeled psychotic, if you keep them out of the area that their psychosis is focused on, they can seem normal." This observation would become the linchpin of Rosenhan's work — that crazy people didn't act crazy all the time; that there was a continuum of behavior that ran from "normal" to "abnormal" within all of us. We all slide around it at various times in our lives, and context often shapes the way we interpret these behaviors.

Under the harsh glare of the hospital's lights, Bill couldn't help but reexamine his own idiosyncrasies, like his tendency to make loose associations and veer into tangents. "When people talk about something it reminds me of something extraneous, and . . . I often bring that into the conversation," he said. "But taken to the extreme, you end up with the clang associations that you get with [serious mental illness]. There's a dividing line in there somewhere. You could probably argue that everyone has something odd. I mean what is normal, what is sane?"

Bill's friend Samson joined him in the step-down unit. He and a few other patients had Bill pegged for a journalist because of his constant writing. "I don't believe you're a real patient. I think you're checking up on the doctors," Samson would say, reflecting a suspicion that Rosenhan also encountered. But not one of the doctors caught on, Bill told me.

One morning, a nurse woke Bill up with a

280

start. "Wake up, Mr. Dickson, you have to go see a doctor. You have diabetes."

Bill was shocked. He'd never had medical issues before — he'd hardly ever had a fever, let alone diabetes. How could he be so sick and no one told him? As he walked with the nurse to the doctor's office, he remembered that his uncle had diabetes and had suffered debilitating side effects. The realization that Bill now had it, too, was chilling — especially when the nurse seemed so nonchalant about it. He'd have to make arrangements to get out as soon as possible to see a doctor; he'd have to tell his wife; he'd have to take those shots every day. Lost in thought, he hardly noticed when the nurse returned and told him he could leave.

"You're the wrong guy," she said. She didn't seem embarrassed or even apologetic. He was simply the wrong guy. Apparently, there was another Dickson on the ward (who was a good deal older, looked nothing like Bill, and lived in another building). The breeziness of the hospital's mistake unnerved Bill. "I mean, jeez, if I was this close to getting treated for diabetes, what if it had been, you know, a lobotomy?"

Maryon visited as often as she could, juggling the kids and the chores while ignoring the chorus of neighborhood questions about her missing husband. She couldn't relax. "I guess I'd seen enough movies or something to know that they could haul him off and, you know, do a brain . . . ," she began, and then stopped. Even from the safe distance of nearly half a century, it was still hard for her to finish. "That they could do a lobotomy."

She wasn't exactly being dramatic. Bad things could and did happen. Bill did not know this, but a psychiatrist who worked at Agnews at the time was nicknamed "Dr. Sparky" by the staff because of his fondness for electroshock therapy. "He would do [it] on anybody — and that includes the staff — if he had the chance," former Agnews social worker Jo Gampon told me. Electroshock started with Italian doctor Ugo Cerletti, who came up with the idea after his assistant visited a Roman slaughterhouse and witnessed how subdued the pigs became after they were shocked with electrical prods on the way to be slaughtered. Oddly, a lightbulb went off. Electroshock took off in America in the 1940s, and Agnews zealously embraced the procedure. A psych technician from that era shuddered when he recalled the weekly lineups. "Our job was to hold the bodies down," he told me. "One, after the other, after the other."

I saw an electroshock box at Patton State Hospital's History of Psychiatry Museum and was pretty surprised at how small and portable the machine looked. This cute machine could do all that? I thought of the movie *The Snake Pit,* when Olivia de Havilland seizes on the table, her head thrashing back and forth, her body stiffening — it turned out the filmmakers did a good job of portraying the procedure, I learned. Patients would sometimes break their backs or necks during the induced seizures. Some would bite straight through their tongues. The "clever little procedure," Ken Kesey wrote in *One Flew Over the Cuckoo's Nest,* "might be said to do the work of the sleeping pill, the electric chair and the torture rack."

Doctors tell me that the treatment today, now called electroconvulsive therapy (ECT), has little in common with the electroshock therapy that Kesey described. ECT is deployed today for patients who are "treatment resistant," the third of people with depression who don't respond to meds. Psychiatrists say that it has evolved "to the point that it is now a fully safe and painless procedure" and is paired with an immobilizing agent to temper any body movements and with general anesthesia so that the patient is unconscious for the duration of the procedure. The amount of current administered is far less than it was then — and memory impairments are reportedly minimal. In one study, 65 percent of patients reported that getting ECT was no worse than going to the dentist. Still, a vocal community, who often picket at APA meetings, say that the possible side effects, including memory loss and cognitive defects, make it "a crime against humanity." In recent years, more hospitals have used it on the East Coast than the West — a product, some say, of Hollywood's vilification of the procedure.

Maryon smuggled in for Bill a copy of the book *Soul on Ice,* a collection of essays written by Eldridge Cleaver, who, while an inmate in a maximum-security prison, chronicled his awakening from a drug dealer and a rapist to a Black Panther and a Marxist.

One of the attendants saw Bill reading the book and struck up a conversation, as if seeing Bill as a human being for the first time.

"What did you talk about?" I asked.

"Well, just about the book, and just about stuff, you know, life in general, women."

"That's interesting, because I haven't heard much about any interactions with the ward, with the attendants. But it seems like it was pretty positive, he treated you like . . ."

"Yeah, yeah, he treated me like a person, in fact he said as he was leaving to go to something else, he said, well, 'You probably won't be around here long,' which I took to mean you're kind of normal so you'll be getting out of here."

As Rosenhan had valued his original, respectful conversation with the attendant Harris (before Harris learned that Rosenhan was a patient, not a doctor), Bill found this interaction gratifying, precisely because it was so rare. He missed being treated as a normal human. He decided it was time to leave.

The how of his release is fuzzy. Rosenhan didn't write anything in his book about it, just that after eight days Bill "suddenly" remembered that he had an event that he had to attend. Bill said he just told the hospital that he wanted to leave (he really wanted to attend a motocross off-road racing event north of San Francisco), and they let him go. There is no indication that he even left with a discharge plan or against medical advice, as Rosenhan said all of the patients did. Did his psychiatrists use the term *in remission*? Did they arrange for him to take meds on the outside or set him up with a support system in the community? Bill didn't think so. I tried to track down the hospital records, but all that remained was one sheet of paper with the "reason for discharge" blank.

One psychiatrist did pull Bill aside, though, and say, "Sometimes, you know, things just kind of seem to build up on people and it's just hard to deal with, and it's really tragic if people do something when they're feeling under that kind of stress that can't be undone."

Bill appreciated the sentiment — clearly, even as Bill was being released, the doctor worried that he was not yet cured, that Bill might become suicidal, and took an extra effort to offer some wisdom on the way out. A day later, he was discharged. He spent nine days in Agnews. That was ten days less than the norm for Rosenhan's pseudopatients, and also a good deal less than Agnews's average stay just four years before, which hovered around 130 days.

In the years that followed his hospitalization, Bill did some informal guest lecturing with Rosenhan at several schools around the country, showing that Rosenhan wasn't quite as careful about keeping identities anonymous as I thought. The resulting fanfare amused Bill, but he was never

Name DICKSON, WILLIAM   Committed as, or
     (Surname)   (First)   (Second)   AKA:
Marital Status: Single
Birthdate: 7-30-44   Birthplace: Texas
Religion: NONE   Spouse: ---
Father: Wilburn Dickson   Mother: Maureen Bird Dickson

         Type of
Case No. Commitm't County Admission Discharge Reason for Discharge
115 733 VOL   SCLA   Nov.12'70   Nov. 20'70

287

tempted to steal any of the spotlight for himself. As time passed, Bill's experience faded into just another seldom-mentioned story from his California days, one that had remained largely unexamined in his own life. When I approached Bill's daughter about the study, she had no knowledge that he had even participated in it.

Bill was one of the final few psychiatric patients to enter Agnews. Two years before, Agnews had begun aggressively rebranding itself as a facility for people with developmental disabilities — and eventually discharged its psychiatric patients, some of whom had been there for decades, to the community or to Northern California's remaining large-scale state institution, Napa State. In 2009, Agnews closed for good, leaving the whole of California with six state psychiatric hospitals, five of which are dedicated solely to housing forensic (criminal) patients.

All that remains of Agnews today is a small, one-room museum on the manicured grounds of software giant Oracle and a sign on the freeway advertising the exit for AGNEWS DEVELOPMENTAL HOSPITAL CENTER, a place that no longer exists.

17
ROSEMARY KENNEDY

The uproar that followed the publication of "On Being Sane in Insane Places" left America with the urgent question: What to do about it? David Rosenhan, Bill Underwood, and the others had provided *Science*-sanctioned evidence for what the anti-psychiatry movement and their ilk had long been arguing: that mental hospitals were relics from a primitive era and should be shuttered. "The anti-psychiatrists could now claim their case was proved. A scientific study in the premier journal of the scientific community had shown psychiatrists did not know the sane from the insane . . . Worse still, the experience of segregation, powerlessness, depersonalization, mortification, and dehumanization . . . were enough to drive a normal person insane," wrote Rael Jean Isaac and Virginia Armat in *Madness in the Streets*.

Experts in and out of the field argued that

these hospitals were "superfluous" institutions, sites of "therapeutic tyranny," and "merely a symptom of an outdated system that is crying for a complete remodeling" that should be "liquidated as rapidly as can be done." By 1973, the year *Science* published "On Being Sane in Insane Places," California governor Ronald Reagan closed Modesto, Dewitt, and Mendocino State Hospitals, converted Agnews into an institution for people with developmental disabilities (a transition that Bill was caught up in), and announced plans to phase out all of the state's public psychiatric hospitals by 1982. One lawyer summed up the view of the time, arguing in 1974 that patients were "better off outside of a hospital with no care than they are inside with no care."

Though Rosenhan and his study likely did more to cement public opinion against psychiatric hospitals than any other academic study, the process to close these institutions had started decades earlier — most significantly with the birth of John F. Kennedy's sister Rosemary.

Rosemary Kennedy's first hours on earth were unimaginable. The doctor was late when her mother's water broke. To slow the delivery until he arrived, the nurse told Rose Kennedy to hold her legs together, and when that wouldn't stop the birth, the nurse pushed the baby's head back into Rose's birth canal, restricting oxygen to the newborn's brain.

From early on it was clear Rosemary was not like her siblings. She had difficulty holding a spoon, riding a bike, and, later, reading and writing. In a family as ambitious as the Kennedys, this made Rosemary a liability. Patriarch Joe Kennedy did his best to keep Rosemary's condition — the official label was "mentally retarded" — out of the public eye. In society pictures taken of Rosemary there is often a family member with her. In one picture, her father clutches her arm as if to keep her physically restrained. But as Rosemary grew, so did her beauty. She was the most attractive of the Kennedy girls, with a curvy figure, beautiful curlicue hair, a love for fine and shimmering clothing, and a wildly captivating full-faced smile.

They sent her off to various schools, where she eventually learned to read at a fourth-grade level. Over time her gregariousness turned sour. Rosemary started to lash out. While living at a convent, she would disappear for hours in the middle of the night. Whatever she did on those walks was a direct threat to the livelihood of the emerging Kennedy dynasty. If she were to, say, get spotted by the gossip columns or, God forbid, get knocked up, this devout Catholic family would never recover. As her behavior grew more unmanageable, Joe Sr. looked for options outside the convent.

His search led him to two American doctors, Walter Freeman and surgeon James Watts, who had imported the lobotomy from Portuguese neurologist António Egas Moniz. Moniz, who received a Nobel Prize for his work in 1949, was inspired to try radical frontal lobe surgery after reading about two Yale physiologists' experiments on chimpanzees. Moniz tested the surgery on humans — severely depressed patients and those with chronic schizophrenia. The procedure, which severed the connections between the prefrontal cortex and the rest of the brain, rendered the patient cured (if by "cured" you meant incontinent and zombified, but easier to manage). Neurologist Freeman would adapt this grueling psychosurgery into a far easier in-and-out procedure nicknamed the ice-pick method, which involved stupefying a patient with rounds of electroshock therapy and then inserting a surgical instrument through the eye socket and — swish, swish, swish — in a few minutes scrambled the structure of the brain.

Lobotomies were not intended for someone with Rosemary's impairment, but that was no impediment. Lobotomies were used to treat everything from homosexuality to nymphomania to drug addiction, all of madness lumped together and treated with one simple surgery, a callback to the early nineteenth-century theory of unitary psychosis. Sixty percent of lobotomies were conducted on women (one study in Europe found that 84 percent of lobotomies were conducted on female patients), even though women made up a smaller segment of the psychiatric population in state hospitals.

Rosemary's sister Kathleen "Kick" Kennedy, a journalist, looked into the surgery and told her mother, "It's nothing we want done for Rosie." It's unclear if Kick's conclusion ever reached her father, however, because he went ahead and booked an appointment with Dr. Freeman and Dr. Watts at George Washington University Hospital in 1941. Rosemary was only twenty-three years old.

The doctors kept detailed notes on their surgeries, so we have a good sense of what Rosemary endured. Dr. Watts drilled burr holes on either side of Rosemary's head, near her temples, and made an incision big enough to allow a small, anvil-shaped instrument — something that looked disarmingly like a tool a bartender would use to make a cocktail — to enter her prefrontal lobe, the most forward part of the brain, associated with higher executive functioning, decision making, and planning for the future. As Rosemary recited a poem or sang a song to communicate her level of cognition, Dr. Watts swiveled the instrument back and forth, back and forth. By the fourth swing, Rosemary became incoherent.

The procedure was an abomination. When she left George Washington University, Rosemary was unable to walk or talk. Only after months of therapy did she regain even the most basic movements. One leg remained forever pigeon-toed, making it nearly impossible for her to get around without help. She communicated with garbled sounds and later graduated to just a handful of simple words. She was like a stroke victim, "a painting that had been brutally slashed so it was scarcely recognizable. She had regressed into an infantlike state, mumbling a few words, sitting for hours staring at the walls, only traces left of the young woman she had been," wrote journalist Laurence Leamer in *The Kennedy*

Women. This was a vivacious, high-spirited young woman who loved beautiful clothes and dancing, who could charm almost anyone. Her mother was so disturbed by the change, one biographer wrote, that it's possible she didn't visit her daughter for more than twenty years. Eventually the family moved her to a private one-story brick home at St. Coletta School for Exceptional Children in Jefferson, Wisconsin, a convent run by Franciscan nuns, where she remained until her death in 2005 at eighty-six. The treatment of Rosemary Kennedy would remain a stain on the family. Rose would later say that what they did to Rosemary was the first of the many tragedies to strike her family, and it brought "yet more danger, death, and sorrow to the Kennedy household."

Rosemary and the "care" she received at one of the most esteemed hospitals in our country made a deep impression on her brother Jack, the future president. In February 1963, eight months before he was assassinated in Dallas, President Kennedy announced: "I have sent to the Congress today a series of proposals to help fight mental illness and mental retardation. These two afflictions have been long neglected. They occur more frequently, affect more people, require more prolonged treatment, and cause more individual and family suffering than any other condition in American life. It has been tolerated too long. It has troubled our national conscience, but only as a problem unpleasant to mention, easy to postpone, and despairing of solution. The time has come for a great national effort. New medical, scientific, and social tools and insights are now available."

His goal was to "get people out of state custodial institutions and back into their communities and homes, without hardship or danger."

In place of monolithic psychiatric hospitals, JFK made a federal commitment to create a network of community-based psychiatric facilities that allowed people with serious mental illness to live outside asylums. This was based on the emerging community psychiatry theories, a reaction to the darkest hour of modern history. "U.S. Army psychiatrists in World War II had observed that chronic war neurosis (today we might call it post-traumatic stress disorder) could be avoided if soldiers were treated in field hospitals just behind the lines, where they could stay in close touch with their buddies and from which they could be discharged rapidly to rejoin their units," wrote Dr. Paul

Appelbaum in his book *Almost a Revolution.* Likewise, community psychiatrists wanted patients to leave state hospitals and stay (for brief periods of time) in acute units, while long-term patients would be released into the arms of the public. Research that showed how "prolonged hospital stays might themselves have negative effects on patients, rendering them 'institutionalized' " — words that Ward 11 staff, Soteria House founders, and Rosenhan himself would have relished — only supported their push. In addition, the new drugs made it possible to imagine a world where the most severely ill could take their meds and live full lives outside the hospital.

"It should be possible, within a decade or two, to reduce the number of patients in mental institutions by 50 percent or more." With this pronouncement JFK signed into law the 1963 Community Mental Health Act, one of the first steps in the phasing out of psychiatric hospitals, which launched "an ongoing exodus of biblical proportions."

Fifty percent or more. It seemed wildly idealistic, but it was conservative considering what would actually happen.

President Lyndon Johnson followed JFK's measure by signing a bill that led to the creation of Medicare and Medicaid in 1965

— federal health care insurance coverage for the poor and elderly — and assigning the federal government the role of "payer, insurer, and regulator" of mental health services. A Medicaid caveat in the form of the Institutions for Mental Diseases (IMD) exclusion prohibited the use of federal Medicaid dollars to fund psychiatric facilities with more than sixteen beds, meaning that most state hospitals, which were almost always larger than sixteen beds, would not receive federal funding. States, realizing that they could transfer the costs of care to the federal government if they closed their hospitals (and if they didn't would be left shouldering the burden of the sickest), began discharging and closing hospitals at unprecedented clips, leaving the mentally ill to vie for limited beds in psychiatric units of general hospitals or the sickest and oldest to nursing homes, which Medicaid covered. The IMD exclusion is still intact and Medicaid continues to be the United States' largest funding source for mental health care. This all funneled individuals with serious mental illness into more " 'medicalized' treatment settings" (like overtaxed emergency departments) and introduced a trend of privatization of quality mental health care that continues today. Despite the passage of a federal mental health parity law in 2008, insurance companies now reimburse mental health professionals eighty-three cents for every dollar covered for primary care, and just over half of psychiatrists take insurance (compared with 89 percent of the rest of medical professionals).

At the same time, civil rights lawyers filed lawsuits against hospitals in the name of human rights. The Bazelon Center for Mental Health Law, founded by a group of lawyers and mental health professionals committed to the rights of people with mental disabilities, opened in Washington, DC, in 1972. Patients, who previously had no representation or recourse (remember when Rosenhan was forced to sign away his rights to be committed?), now had an army of lawyers working to keep them out of hospitals, or to help them get discharged as quickly as possible. A series of landmark acts, including the Lanterman-Petris-Short (LPS) Act that Bill encountered during his stay, pushed for patients to be treated in the "least restrictive settings" with minimum patient-to-staff ratios. Stricter commitment laws required patients to either be "gravely disabled" or pose an imminent threat to themselves or others to be committed. Vague notions of voices that say "thud, empty, hollow" certainly would no longer do it. There was the 1971 *Wyatt v. Stickney* ruling, which said that if the state was unable to meet the standards of minimally required care, you could not force hospitalization. In response, the hospitals didn't revamp and update. They closed.

Because, most opportunistically, shuttering these institutions would save some serious cash, making everyone along the political spectrum happy. The "mental illness treatment system had been essentially beheaded," wrote psychiatrist E. Fuller Torrey.

And as Bill's and Rosenhan's hospitalizations and Rosemary Kennedy's lobotomy proved, good riddance. Right?

JFK would not live long enough to see his work's aftermath. From the year of his death, 1963, to the publication of Rosenhan's study in 1973, the total resident population in state and county psychiatric hospitals dropped by almost 50 percent, from 504,600 to 255,000. Ten years later, the US psychiatric population would drop another 50 percent to 132,164. Today 90 percent of the beds available when JFK made his speech have closed as the country's population has nearly doubled.

Trouble is, for all of its idealism and promise, the dreams of community care were never actualized because the funds never materialized. The money was intended to follow the patients. It didn't. The community care model at its very best provided nominal care to the least impaired. Those with the most severe forms of these disorders were ignored or cast aside. The new community facilities themselves actually resembled "small long-term state hospital wards," wrote Richard Lamb as early as 1969. "One is overcome by the depressing atmosphere."

The government policies that closed these institutions did not embed people more deeply into the community — they pushed them further outside onto our streets and into our homeless shelters, even, as we'll see, into our prisons.

As one psychologist, who practices today in the shadow of deinstitutionalization at a forensic psychiatric facility, told me: "We could see the light at the end of the tunnel. We didn't know that it was an oncoming train."

■ ■ ■ ■

PART FOUR

■ ■ ■ ■

When the going gets weird, the weird turn
professional.
> — Hunter S. Thompson, "Fear and
> Loathing at the Super Bowl"

18
THE TRUTH SEEKER

I have no doubt Rosenhan would have been pleased that his study played a role in the closing of these institutions. In a letter written a few days after "On Being Sane in Insane Places" was published, he corresponded with a psychiatrist who had suggested that the study could be interpreted in a different way: Perhaps *more* money should be allocated to these institutions to improve them? Rosenhan did not agree: "I'm simply not sure that more money in this area is going to help and indeed, sometimes I wonder whether less money would be better for the patients."

Rosenhan had been so certain about his convictions. Certainty was a luxury that I could no longer afford. The deeper I looked, the more complicated the story grew.

This new uncertainty came from aspects of Bill's story that had bothered me. Throughout Rosenhan's notes, I kept running into sloppiness that seemed unprofessional and possibly unethical — mistakes made about length of time spent in the hospital (it was minor, but Bill had spent eight days in, while Rosenhan had written repeatedly that Bill had spent seven), wildly inaccurate patient numbers (Rosenhan had written that "Bill Dixon's" hospital held 8,000 patients, while there were only 1,510); he even misspelled Bill's pseudonym in his private notes, using Dixon rather than Dickson (though this may have been on purpose). There were also discrepancies between what Rosenhan wrote and what Bill remembered: Bill wasn't released with his diagnosis "in remission," while Rosenhan wrote that all the pseudopatients had been. Bill also did not recall recording detailed data notes, like the number of minutes the staff spent on the ward — highly specific numbers included in the early drafts and the published paper. Rosenhan listed percentages of how psychiatrists and nurses behaved on the wards when faced with a pseudopatient (71 percent of psychiatrists moved on, head averted, while 2 percent paused and chatted, for example). Rosenhan also wrote that attendants spent an average of 11.3 percent of their time outside the cage and on the floor, while nurses emerged an average of 11.5 times per shift. "He certainly wouldn't have gotten exact numbers from me because I didn't really watch the office that closely. I just told him how often I had seen nurses/attendants out and about on the ward," Bill told me. If Bill, a graduate student studying psychology, didn't gather this information, who had?

It bothered me that Rosenhan had told Maryon he had procured writs of habeas corpus when he hadn't. I didn't like how blithely Rosenhan had sent Bill in and how little he had prepped him, which resulted in Bill ingesting a large dose of Thorazine. Had Rosenhan learned nothing from the six other pseudopatients he had trained before Bill? Similarly bothersome, Rosenhan didn't fully vet Agnews, which was in a state of disarray as it prepared to close its doors — a dangerous and unfair time to send someone inside for the experiment. The uniquely chaotic transition occurring then at Agnews should have disqualified it because the results would hardly be generalizable.

Rosenhan had taken great efforts to ensure his own safety when he went undercover, alerting the superintendent and even requesting a tour of the hospital prior to his stay. But for his student, there is no indication that these precautions were taken.

Wasn't it his duty as a researcher, as a teacher, and most of all as a human, to make sure that Bill was properly equipped for a traumatic and possibly dangerous experience? It didn't sound like the Rosenhan I had come to know through his writing and my research. This didn't just make me question Rosenhan's character, it also undermined the study. It was key that Rosenhan had limited the amount of variability in the presentation of their symptoms (voices that said "thud, empty, hollow") to make the data *mean something.* Not preparing his pseudopatients adequately harmed the study's validity.

Still — there was no guarantee that Bill remembered everything accurately, which could account for some of the inconsistencies — so I revisited the CRITICISM folder, located in Rosenhan's private files, hoping that some insight would jump out at me from the chorus of hostile voices:

- "Seriously flawed by methodological inadequacies." — Paul R. Fleischmann, Department of Psychiatry, Yale University
- "It appears that the pseudopatient gathered pseudo data for a pseudo research study . . ." — Otto F. Thaler,

Department of Psychiatry, School of Medicine, University of Rochester
- "If I were to drink a quart of blood and, concealing what I had done, come to the emergency room of any hospital vomiting blood, the behavior of the staff would be quite predictable. If they labeled and treated me as having a bleeding ulcer, I doubt that I could argue convincingly that medical science does not know how to diagnose the condition." — Seymour Kety, McLean psychiatrist, who studied the genetics of schizophrenia
- "To point out that Rosenhan's conclusion is unwarranted on the basis of his, ah, data is perhaps belaboring the obvious . . . Why did Science publish this?" — J. Vance Israel, Medical College of Georgia

Why did Science *publish this?* I had pondered this same question earlier in my research, and had asked *Science* if they could provide any information about the review process prior to the study's publication. Rosenhan couldn't have just mailed in a copy of his article and twiddled his thumbs as the prestigious journal made his career. He would have had to take part in a

309

peer review process; someone on the editorial board would have inquired about his data, about the pseudopatients, about the hospitals. That's how it works — certainly how it's *supposed* to work.

Unfortunately, *Science* wasn't going to give up these answers. A representative said that she would not divulge any details about the process because it was confidential; the journal protected its reviewers. I recruited the help of sociologist Andrew Scull to reach out on my behalf as an academic, but they declined his request for a different reason: They said that they don't keep records that far back. In a letter to a colleague who wanted to publish his own follow-up pseudopatient research, Rosenhan said that he picked *Science* "mainly because they have a very quick review system. It usually takes no more than two months to hear from them, and four or five months for the article to be in print." Psychologist Ben Harris has another theory why Rosenhan submitted his study to *Science.* He thinks that because it is a generalist journal (meaning that it has a wide range of interests beyond just psychiatry, unlike a more specialized journal like *Molecular Psychiatry*), he may have found a back door into academic fame. "Submitting to *Science* [may have been] a trick that [could have] bypassed review by top people in the field of clinical psychology," Harris said.

Because of the stature of the journal in which it was published, none of the intense critiques from inside the field seemed to land — not really. Psychiatrists were like hungry panthers pouncing on a prey that had strayed too far from its pack, a prey (a psychologist, which was even worse) that had preened and boasted and received more attention than most of them ever would. The lay public, who were already primed to be suspicious of the field thanks in part to the mounting anti-psychiatry movement, were hardly inclined to be sympathetic to disgruntled psychiatrists with their reputations on the line. The more the psychiatrists had gnashed their teeth, the stronger the study's power had grown.

Still, one criticism seemed to unmoor Rosenhan. I know this because he kept five copies of this critique in his files, despite, I remind you, keeping none of the pseudopatient notes. The article "On Pseudoscience in Science" was written by Robert Spitzer, the man who helped remove the term *homosexuality* from the *DSM-II*. The piece is delicious in its biting bitchiness. It's the drollest piece of academic literature I've ever read.

311

It's mean. It's funny. And man does it pack a hell of a wallop.

"Some foods taste delicious but leave a bad aftertaste," Spitzer began. "So it is with Rosenhan's study, which by virtue of the prestige and wide distribution of *Science,* the journal in which it appeared, provoked a furor in the scientific community." He called the paper "pseudoscience presented as science" and wrote that its conclusion "leads to a diagnosis of 'logic in remission.' " Spitzer then tore into every aspect of the Rosenhan paper — "one hardly knows where to begin" — from his research methods, which he called "unscientific," to his use of the terms "sanity and insanity,"[1] which are legal concepts, not psychiatric diagnoses. (Rosenhan defended his use of the terms in a letter to Vermont psychiatrist Alexander Nies in 1973: "Sane comes closest to what we mean when we

1. *Insanity* in a legal context involves intent — it's a question of whether or not the defendant was able to determine right from wrong during the crime. Here is the definition from Law.com: "n. mental illness of such a severe nature that a person cannot distinguish fantasy from reality, cannot conduct her/his affairs due to psychosis, or is subject to uncontrollable impulsive behavior."

say 'normal' (just imagine the fuss over that word)."

Spitzer argued that the designation "in remission," a term rarely used but applied to all eight of the pseudopatients (though, it seems, not to Bill), actually showed that the doctors *were* aware that these pseudopatients were different from the rest. He called Rosenhan out for failing to disclose his data and his sources. Spitzer implied that Rosenhan was willfully withholding information from readers. "Until now, I have assumed that the pseudopatients presented only one symptom of psychiatric disorder. Actually we know very little about how the pseudopatients presented themselves. What did the pseudopatients say in the study reported in *Science,* when asked, as they must have been, what effect the hallucinations were having on their lives and why they were seeking admission to the hospital?" Spitzer asked.

Rosenhan took particular umbrage at Spitzer's assertion that Rosenhan refused to share his and his pseudopatients' medical records. I know this thanks to another folder titled SPITZER, ROBERT, which held a series of heated private letters between the two men.

Rosenhan and Spitzer began corresponding a year after "On Being Sane in Insane Places" was published, when Spitzer, while in the middle of writing his critique, was helping to arrange a symposium on Rosenhan's study sponsored by the *Journal of Abnormal Psychology.*

The first letter opened "Dear Dave," which struck me as odd because Rosenhan didn't often go by Dave. It was a false familiarity that feels more like an elbow to the rib than a handshake. Spitzer began by cordially asking Rosenhan for a list of references that cited Rosenhan's study. A close reading of Rosenhan's response, however, reveals an undercurrent of rage. I imagine Rosenhan sitting among piles of papers on his desk, his forefinger at his temple, reading this missive, his face growing redder and redder; while I imagine Spitzer gleefully typing up his pages, smiling to himself as he thought of a zinger, maybe even editing his words to sharpen the shiv so that it pierced right into the heart of the paper's shortcomings.

Spitzer himself had been long obsessed with hard data and classification. There were stories that as a boy attending sleep-away camp, he designed a rating scale to track the hotness of his fellow female campers. By his teens, he had developed an active interest in psychoanalysis, specifically Reichian psychology and its orgone box therapy,[2] a sham treatment fad popular in the 1940s and 1950s that claimed to use universe energy to ease psychic illness (and also espoused a belief in extraterrestrials). Spitzer subjected the orgone box to a series of experiments and found that the box was just that, a box, and had no effect whatsoever on the person inside it. This study was completed before Spitzer could legally drink alcohol.

A quieter motivation came from a strain of deep unhappiness that shadowed his family. Spitzer's grandfather had pitched his own wheelchair out of a window after being struck by a neurological illness. His mother struggled with depression, the illness culminating after his older sister passed away

2. Another Reichian with a rumored orgone box of his own was Vermont senator Bernie Sanders. In 1969, he wrote an essay called "Cancer, Disease and Society" for the *Freeman,* quoting from Wilhelm Reich's 1948 book *The Cancer Biopathy,* writing, as *Mother Jones* reported, that he was " 'very definite about the link between emotional and sexual health, and cancer,' and he walked readers through Reich's theory about the consequences of suppressing 'biosexual excitation.' "

from encephalitis when Spitzer was just four years old. Despite appearances to the contrary, Spitzer, a passionate, forceful, and animated man, inherited the family darkness. He struggled with depression and feelings of worthlessness and would spend his career comfortable with the solidity of numbers and hard facts.

Spitzer was, above all else, "a truth seeker," his wife, Janet Williams, told me, and Rosenhan's study piqued his intellectual interest.

In their correspondences, the two men traded passive aggressive attacks, shaded with affability — each ended his missive with "Yours Sincerely" (Rosenhan) or "Sincerely yours" (Spitzer) — back and forth. Spitzer repeatedly asked for access to the other pseudopatient materials and Rosenhan sidestepped him, explaining that the files contain sensitive information. When Spitzer wouldn't remove a statement that Rosenhan "refuses to identify" the hospitals, Rosenhan got defensive: "[This] implies that I have something to conceal. You know that is not the case. Because my study has been misinterpreted to suggest that psychiatrists and hospitals generally, are incompetent, I am obliged to protect these sources," wrote Rosenhan. (After publication, Rosenhan began to pull back on some of his harsh criticism by soft-pedaling some of his paper's conclusions. "Let me make clear," he wrote in a response letter to his critics published in *Science,* "that the theory that underlies this effort, and the report itself, do not support the vilification of psychiatric care.")

And then Rosenhan launched his own attack: "In the same vein, I offer some observations about your own paper. Both the title and the abstract contain the phrase 'pseudo-science in science'. That phrase is needlessly pejorative. What is pseudoscience other than findings that one disagrees with? Does science, in your view, have a particular method, or guarantee particular findings? Especially as you agree with a number of the findings, there must be other ways of indicating that you disagree with some methods and interpretations without treading such thin ice. 'Logic in remission', also in the title and the abstract, is a personal remark. Your argument can be strengthened considerably by dealing with the paper — its logic, in your view, is faulty — rather than its author."

Spitzer returned with some critiques of his own and thumbed his nose at Rosenhan's statistical interpretation of data. "Perhaps all that we can hope for is that our letters to each other get progressively shorter," Spitzer quipped.

From here on, Rosenhan's writing is the angriest I've ever seen; he's practically spitting. He recruited Loren Mosher (founder of Soteria House) for his advice and even asked Haverford Hospital superintendent Jack Kremens to reach out to Spitzer on his behalf to convince him not to publish his critique. His argument was that the hospital would suffer a needless stain on its reputation. And he added: "You now have it from myself and the superintendent of the hospital (who arranged my hospitalization) that my stay there was part of a teaching exercise and had nothing directly to do with research."

Wait, wait, wait.

Haverford Hospital had nothing to do with his research? A mere teaching exercise? Sure, it may have started that way, but Rosenhan couldn't reasonably argue that he did not include his Haverford stay in "On Being Sane in Insane Places." Most, if not all, of the in-depth scenes in the study are about Rosenhan's hospitalization. When a patient comes up to a pseudopatient and says, "You're not crazy. You're a journalist or a professor. You're checking up on the hospital," it is taken verbatim from Rosen-

318

han's notes on the ward. Rosenhan was the one who watched as a nurse adjusted her bra in front of patients. He even quoted directly from the medical record written by Dr. Bartlett, the doctor who committed him. How could he possibly say that Haverford was just a test run?

That was an outright lie. And Rosenhan knew it.

Rosenhan knew it, and Spitzer knew it, too. The truth seeker had managed to gain access to Rosenhan's medical records, the same pages I myself had tracked down. The pages I now held in my hands.

19
"ALL OTHER QUESTIONS FOLLOW FROM THAT"

In therapy, the aha moment is the stage of realization when sudden clarity hits and feelings that you have suppressed come to the fore and begin clicking into place. Robert Spitzer offered this to me from a distance of four decades.

I dug into the medical records. On cursory reading, the records support Rosenhan's paper: There was his pseudonym David Lurie; there were the accurate numbers of days he spent hospitalized (though I had noticed that sometimes he exaggerated this figure depending on the audience); and there were his diagnoses, "schizophrenia, schizoaffective type," and later, "paranoid schizophrenia, in remission." It conformed to his published paper. It checked out.

Except it didn't, as Spitzer had found.

One of the foundational principles of "On Being Sane in Insane Places" was that all of the pseudopatients presented with *one symptom,* voices that said "thud, empty, hollow." The only other amendments were meant to add a layer of protection for the participants, changing names, jobs, addresses, but "no further alterations of person, history, or circumstances were made," Rosenhan wrote.

But this is immediately contradicted by the text of the intake interview, written by Dr. Bartlett, the man who first diagnosed Rosenhan and insisted that Mollie commit him. If Dr. Bartlett's notes are to be believed, Rosenhan's alleged symptoms went far beyond "thud, empty, hollow."

This is what Dr. Bartlett recorded:

ADMISSION NOTE: 2/6/69
The patient is a 39 year old married father of two, living with his wife. 3–4 months ago he started hearing noises, then voices. Recently he has been able to discern that the voices say, "It's empty", "nothing inside". "It's hollow, it makes an empty noise." He has felt that he is "sensitive to radio signals and hear what people are thinking." He realized that these experiences are unreal but cannot accept their reality. He has realized that these insulate out the noises by putting "copper over my ears". One reason for coming to the hospi-

tal was because things "are better insulated in a hospital". He has also had suicidal thoughts.

The first part checks out — again we see the key words *thud, hollow, empty.* But then Rosenhan goes off script. Bartlett wrote that Lurie was so disturbed by the voices that he had to put copper over his ears — an almost clichéd example of the "tinfoil hat delusion" commonly reported by people suffering from serious mental illness.

"He has felt that he is 'sensitive to radio signals and hear[s] what people are thinking.'"

Hallucinations and disturbances in thought patterns, especially the belief in the ability to hear or control other people's thoughts, is considered a key symptom of schizophrenia, one of Kurt Schneider's "first rank symptoms for schizophrenia." In Massachusetts General Hospital's *Handbook of General Hospital Psychiatry,* "thought broadcasting," or the belief that others can hear your thoughts or the thoughts of others, is a classic symptom for a quick and easy identification of psychosis in an emergency room setting. It was the sort of symptom I had displayed myself during my encephalitis when I believed I could read the nurses' thoughts about me, or that I could age people with my mind.

On deeper examination, the red flags continued to wave. There is a philosophy of the psychotic experience underlying Rosenhan's paper that feels authentic. According to Clara Kean, who wrote about her experience with schizophrenia in two articles for *Schizophrenia Bulletin,* psychosis involves an "existential permeability," a belief that there is a softening of the space between the self and others. She described the experience as the "dissolution of ego boundaries," when "what is originated from the self and what is not are confused." I recognize Clara's words in my own experience. When I was psychotic, I became more attuned to my surroundings (even if this attention was distorted, confused, misdirected) while also experiencing a loss of self that felt dangerous, more frightening than any other symptom I experienced. Whether intentionally or not, Rosenhan touched on something real, something that a good psychiatrist would identify as a fairly typical, though traumatic as hell, part of being psychotically ill.

Rosenhan's timeline as reported to the doctor is also much longer than recorded in his paper. Bartlett wrote that Rosenhan started hearing voices more than three months before his admission, and that the hallucinations, in the form of amorphous sounds, started at least six months prior to that. According to another psychiatrist, Rosenhan "dated his illness to *ten years ago* [emphasis mine] when he gave up his job in economics."

All of these factors created a "much clearer picture of schizophrenia, even by today's standards," according to Dr. Michael Meade, the chairman of psychiatry at Santa Clara Valley Health and Hospital System. (Dr. Meade added that it was unlikely David Lurie would have received a schizophrenia diagnosis today, however — the age of onset was too unusual, for example; he would likely have received the no-man's-land diagnosis of "psychotic disorder, not otherwise specified.") Still, the symptoms did conspire to create a realistic portrait of a man suffering from some kind of illness — not merely an "existential psychosis," as Rosenhan said he intended.

In the same intake interview with Dr. Bartlett, Rosenhan also said that Mollie "did not know how disturbed and helpless and useless" he was and that he had "thought of suicide" and believed that "everyone would be better off if he was not around."

STATES: "Mr. Lurie has been hearing voices and noises for last three months which he has tried to insulate out by putting copper over his ears. He has been unable to work or concentrate. Medicine he was given did no good. Mr. David Lurie is a tense, anxious appearing man who had difficulty expressing himself. He hears noises and voices which say, "It's hollow", "everything is empty". He has thought of suicide as everyone would be better off if he was not around."

The thoughts of suicide and threats of self-harm, called suicidal ideation, would provide grounds for immediate and necessary commitment. "Active psychosis is one of the most serious comorbid risk factors in suicidal patients," Dr. Meade said. "To not hospitalize such a patient would be professionally unethical, and, in almost every circumstance, malpractice." No wonder Bartlett was so insistent that Mollie sign the forms. Rosenhan gave them no choice but to commit him.

This seemed pretty damning. Out of fairness, could there be any other explanation? Was it possible that Rosenhan was being honest here, that he was feeling suicidal at the time? Problematic as it would be for him

325

to present himself as a "sane, healthy" control case in a study about mental health if he was also sincerely suicidal, was there any possibility that he was following the rules, if not the spirit, of his own experiment, and telling the truth about everything but the voices?

When I emailed Florence and asked if she knew whether Rosenhan was ever suicidal, she wrote: "It seems to me that any sentient human being, and surely Rosenhan was sentient, has considered suicide." She added that some of his angry outbursts (he didn't lose his temper often, but when he did it was frequently dramatic) might easily have been by-products of undiagnosed depression. But Florence acknowledged that the way the doctors portrayed his suffering was more urgent and potentially unsafe, and she firmly doubted that Rosenhan was ever clinically suicidal. At no time in their close friendship had he ever discussed feelings of desperation that cut this deeply.

Yet, in his intake interview, Rosenhan elaborated with more fabrications — about a long-running feud with an employer and issues with work, adding a layer of desperation that would only heighten his suicide risk. In the interview, Rosenhan mentioned that after he had lost his job in advertising, his wife had to take a part-time job typing and they had to borrow money from his in-laws. "This has been very embarrassing," Dr. Bartlett quoted "David Lurie." Yet as far as I have been able to determine, not a word of this is true.

Furthermore, the two other doctors who examined Lurie not only corroborated Dr. Bartlett's impressions of the patient's mental state, but expanded on them. Dr. Browning wrote that Lurie had "placed the bottom of a copper pot up to his ear to differentiate the noises that he was hearing and he tried to interfere with this signal he thought he was receiving" and that he had contemplated suicide but thus far not taken any action because, as Browning quoted Lurie, "I don't have the guts."

In the most charitable reading possible, one could imagine that perhaps Rosenhan worried that his "thud, empty, hollow" symptoms would not be enough to get himself inside the hospital, so he had exaggerated his story to ensure admittance for what, at the time, had been a mere teaching exercise. (None of this, of course, excuses the choice to use tainted data in the study later on, nor to lie about it to Spitzer in the aftermath.) Or perhaps he felt the curious dynamic so often present in doctor-patient relationships where patients want to impress clinicians or convince them of the legitimacy of their suffering by offering up heightened details. Either way, I could now picture Lurie more accurately from Dr. Bartlett's perspective: a "tense, anxious" middle-aged man whose suffering had grown so acute that he decided to check himself into a psychiatric hospital. What else could Dr. Bartlett have done but help him?

No matter how much benefit of the doubt one might try to give him, clearly the full story wouldn't be found in Rosenhan's papers alone. I had to find Dr. Bartlett.

Unfortunately, it turned out I was almost three decades too late to hear Bartlett's story firsthand. Dr. Frank "Lewis" Bartlett had died on May 24, 1989, at the age of seventy-four. He spent three decades working in mental health care, according to his obituary. I tracked down his surviving daughter, listed as "Mary Bartlett Giese of Chevy Chase, MD."

Dr. Bartlett's interest in psychiatry came from his love for his troubled but beautiful wife, Barbara Blackburn, who became deeply ill shortly after the birth of her first child, Mary's brother Gus. Before he became a psychiatrist, Dr. Bartlett was a rabbit farmer who joined the Merchant Marine, leaving his wife and young child at home. Neighbors intervened when they discovered that Gus, just three years old, had been left to fend for himself while his mother refused to leave her bed for several weeks. This led to Barbara's first psychiatric hospitalization in California. When she returned home, she spiraled into a depressive state so severe that her own son found her in the kitchen with her head in the oven ready to end her life, at which point her husband gave up rabbit farming, enrolled in medical school, and moved the family to Vermont.

Bartlett became obsessed with finding a cure for his wife, even after she eloped to California with a fellow psychiatric patient, leaving Dr. Bartlett to raise their two children alone. He published passionate op-eds decrying the treatment of the mentally ill in America and coined the term *institutional peonage,* comparing forced work during hospitalization to slavery. He even began a pen-pal relationship with Ken Kesey after reading *One Flew Over the Cuckoo's Nest,* admitting in one mournful letter that Kesey's use of lobotomy in the novel's climax gave him a "creepy feeling" as he remembered "two young colored girls I worked up for a lobotomy ten years ago."

Until the very end, deep into retirement, even after the cigarettes got the better of his lungs, these issues still dominated his life. He formed a small group called the Philadelphia Advocates for the Mentally Disabled, basically a helpline that you could call at any hour and Bartlett or one of his associates would come and help a psychotic person on the street find a safe and warm place to stay the night. At his funeral a close friend said, "I just have this picture of Lew coming down the street in that old Plymouth, and it's snowing, and he's talking to some guy in a box. And eventually the man emerges and agrees to go to a shelter."

When I told Mary about Rosenhan's study and about Dr. Bartlett's miscalculation, she told me that he never discussed it with her (and since he was never named, his role in it never became public), but she was sure that it had "hurt him deeply." This Dr. Bartlett, a man whom I — and likely many of Rosenhan's readers — had first imagined as a bumbling stereotype, had lived a life dedicated to the cause, a man who intimately understood the toll serious mental illness takes on a person and on a family. Dr. Bartlett wasn't a bad doctor who made a bad decision. He wasn't even a good doctor who made a mistake. He was a good doctor who made the best call given the information he received.

If I could get Bartlett so wrong, had I also been reading Rosenhan wrong?

And then there was the interview with Rosenhan's colleague Ervin Staub, emeritus professor of psychology at the University of Massachusetts, Amherst.

Before I continue, remember: Rosenhan was bald. I've mentioned this fact repeatedly because it was one of his most defining characteristics. He lost his hair as a young man, and when people describe him, his domed head and his deep voice are the two features that come up over and over again.

Professor Ervin Staub, like Rosenhan, studies altruistic behavior in children and adults. His key work is on "active bystanders," or the study of the people who witness a situation and do (or do not) offer help. (I'm sure I'm oversimplifying, but Ervin's work reminds me of the *Seinfeld* finale when Elaine, Jerry, George, and Kramer witness a carjacking, do nothing, and are arrested on a "duty to serve" violation.) Rosenhan befriended Ervin when he came to Stanford in 1973 as a visiting professor. At a party at Rosenhan's house (such parties were legendary), Rosenhan regaled a group of people with the story of his hospitalization, mesmerizing the crowd with his dramatic tale. He spoke about how "difficult it was to get out." At one point Rosenhan described a wig that he wore to hide his identity.

"Do you want to see it?" Rosenhan asked.

Rosenhan took Ervin and company upstairs to his bedroom where he kept the wig.

"It was somewhat wild, a bit long," Ervin said. "It was an interesting wig — kind of right for a professor." We both laughed out loud at the thought of Rosenhan hamming it up with a long wig. After a few more questions, I thanked him for the enjoyable interview.

It wasn't until I returned to the medical records that I stopped at his medical care plan. Not only had Dr. Bartlett described a "balding" David Lurie, but there was also a picture attached to his record: In it, Rosenhan stares straight ahead. Though the photocopy is dark, you can still see the gleam reflecting off Rosenhan's hairless head.

Rosenhan wore no wig during his hospitalization.

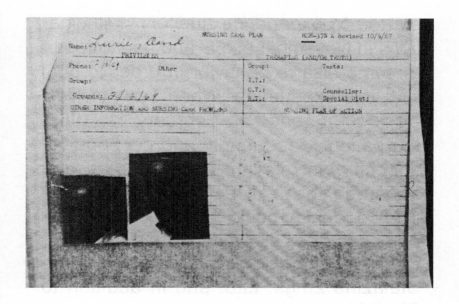

As bewildering as the wig story was on its own, the full extent of his distortions came to light once I placed the published study next to the medical record. Rosenhan had even amended the parts of the medical record that he excerpted in his paper, exaggerating and focusing on certain details while dropping other ones.

This 39 year old, white, married, Jewish male was admitted on February 6, 1969 on 314 commitment. The patient came to the hospital on his own volition and apparently was seeking help. Review of the history reveals that since summer of '68' the patient has stopped working and has shown a definite social withdrawal. He started to experience auditory hallucinations in November of '68' and had to resort to some bizarre behavior in order to deal with this experience. When seen in New Case Conference on February 11, 1969, the patient was friendly and cooperative, speech was relevant and coherant, and appeared to be of extremely high intelligence. Since being hospitalized he reports complete alleviation of his hallucinatory experiences, and had been on Stelazine 2mgs. t.i.d. and in addition to Elavil 25 mgs. t.i.d.

The medical record

This white 39-year-old male . . . manifests a long history of considerable ambivalence in close relationships, which begins in early childhood. A warm relationship with his mother cools during his adolescence. A distant relationship to his father is described as becoming very intense. Affective stability is absent. His attempts to control emotionality with his wife and children are punctuated by angry out-

334

bursts and, in the case of the children, spankings. And while he says that he has several good friends, one senses considerable ambivalence embedded in those relationships also. . . .

The version published in "On Being Sane in Insane Places"

The medical record included no reference to his fluctuating relationship with his parents — nothing about a "warm relationship with his mother" that cooled during his teenage years or a "distant relationship with his father" that intensified with age. Neither of these sentences appeared at any point in his record: "manifests a long history of considerable ambivalence in close relationships, which begins in childhood" or "while he says that he has several good friends, one senses considerable ambivalence embedded in those relationships also." Even though Rosenhan wrote in his published paper and more extensively in his unpublished book that a psychiatrist fixated on a spanking episode involving his son, there is no mention of this in the medical record, either. Rosenhan invented all of this, while conveniently excising any reference to copper pots or suicidal thoughts.

In "On Being Sane in Insane Places," Rosenhan wrote: "The facts of the case were unintentionally distorted by the staff to achieve a consistency with a popular theory of the dynamics of schizophrenic reaction."

Instead, it was becoming alarmingly clear that the facts were distorted *intentionally* — by Rosenhan himself.

What else, then, was misrepresented in Rosenhan's study? I'd only begun getting to the truth after my conversation with Bill; now I understood that the other six pseudopatients were the only ones who could fill out the real story. But I didn't know where to begin looking for them. I didn't know what hospitals they'd been in. I didn't even know their real names.

20
CRITERIONATING

In the back-and-forth between Rosenhan and Spitzer, Rosenhan seemed fixated on how Spitzer managed to get his hands on the records, focusing on this transgression to deflect from his own. Eventually, through sheer force of rage, Rosenhan learned that Spitzer received the records, secondhand, from Haverford State Hospital itself. Dr. Bartlett, feeling slighted by Rosenhan's paper and its misleading portrayal of the care he received, sent Rosenhan's medical records to a psychiatrist named Robert Woodruff, who would later join the *DSM-III* task force. Woodruff was vocal in his critiques of Rosenhan's study and had written a fiery op-ed in the *Medical World News*, which Bartlett had seen. When Dr. Woodruff heard that Spitzer was organizing a conference on Rosenhan's paper, he sent Rosenhan's records to Spitzer. Spitzer knew everything that we know now — how far

Rosenhan exaggerated his symptoms, how he unambiguously exaggerated some of the portrayals of his care — yet Spitzer never published these findings. If Spitzer, the "truth seeker," had all the same information I had, why hadn't he sounded the alarm about this popular study that was embarrassing his profession?

Once again, however, it was too late for me to find out. Woodruff took his own life in 1976, so I could not ask him why he remained silent. By the time I learned about the records, Spitzer was battling serious health problems that restrained him from sparring in the arena of academic controversy. The last time that the public heard from him was his 2012 denouncement of his prior research that supported the use of conversion therapy. And then the day after Christmas in 2015, the *New York Times* ran Spitzer's obit: "Dr. Robert L. Spitzer, who gave psychiatry its first set of rigorous standards to describe mental disorders, providing a framework for diagnosis, research and legal judgments — as well as a lingua franca for the endless social debate over where to draw the line between normal and abnormal behavior — died on Friday in Seattle. He was 83."

I'm left with the actions and words he left

behind. Why had he said, once, that his critique of Rosenhan's study was the paper he was most proud of, "the best thing I have ever written"? Spitzer had even returned to the Rosenhan well in 1976 by writing a follow-up on Rosenhan's study called "More on Pseudoscience in Science and the Case for Psychiatric Diagnosis." In it, Spitzer concluded that despite the paper's glaring issues, Rosenhan got one thing right: his "recognition of the serious problems of the reliability of psychiatric diagnosis" — and Spitzer had a plan in place to solve it.

"For Spitzer, paradoxically, Rosenhan's study and the extraordinary publicity it received was manna from Heaven. It provided the final impetus for a study he had been agitating to conduct for some time, to set up a task force of the American Psychiatric Association charged with revamping psychiatry's approach to diagnosis," wrote sociologist Andrew Scull.

In other words, the study was instrumental in achieving Spitzer's goals: It gave him the grounds to move forward with the overhaul he knew the field needed to survive. So why deliver the fatal blow to something that could be so useful?

In the spring of 1974, APA medical director Melvin Sabshin tapped Spitzer to shepherd the creation of a new version of the *DSM*, setting in motion a "fateful point in the history of the American psychiatric profession." The job was perfect for Spitzer, which worked for everyone because no one else wanted it. Most psychiatrists were far too enamored with sexier, Technicolor explorations of the motivations behind human behavior (with its mining of Greek myths like Oedipus and Electra for sources of interior conflict) to take on the drab black-and-white statistical backwater of diagnosis.

This new manual would be nothing like the *DSM-I,* a puny spiral-bound booklet created in 1952 after physicians witnessed the psychic horror that war wreaked; it would render the *DSM-II,* an analytically oriented text that used Freud-friendly terms like *psychoneurotic* and *phobic neuroses,* obsolete.

This third edition would highlight the teachings of psychiatrists reemerging at that time. "They were determined to create a psychiatry that looked more like the rest of medicine, in which patients were understood to have diseases and in which doctors identified the diseases and then targeted them by treating the body, just as medicine identified and treated cardiac illness, thyroiditis, and diabetes," wrote Tanya Marie

Luhrmann in *Of Two Minds.*

Spitzer recruited from the staunchly anti-Freudian, biologically focused constituency at Washington University in St. Louis a group of like-minded psychiatrists who called themselves neo-Kraepelinians, a direct callback to the German psychiatrist who proposed a new diagnostic language with dementia praecox. The Wash U group also referred to themselves as DOPs, or data-oriented persons, whose "guns [were] pointed" at psychoanalysis. Rumor was that they kept a picture of Freud above the urinal in their bathroom. In 1972 the Wash U contingent published the "Feighner Criteria," one of the most cited papers in modern psychiatric history, which provided rigorous diagnostic criteria based off a descriptive approach — or the grouping symptoms that are common to diagnosis (again, much as Kraepelin did in the late 1800s) — and set the groundwork for Spitzer's *DSM-III.*

In 1980, the third edition of the *Diagnostic and Statistical Manual of Mental Disorders* roared to life. The big fat book (494 pages, compared with the *DSM-II,* which was 134 pages) offered up 265 disorders, more than double the number found in the first edition. The manual scrubbed most psychoanalytic references found in the previous *DSM*s and successfully ushered psychiatry back into the good graces of mainstream medicine. The *DSM-III* introduced "axes." Axis I was devoted to disorders such as anxiety, anorexia, schizophrenia, and major depression. These were different from the personality disorders (like borderline, sociopathic, and narcissistic personality disorders) and developmental disorders in Axis II, described as "conditions and patterns of behavior that are defined as enduring, inflexible, and maladaptive." The third axis was devoted to "physical" disorders, like cirrhosis of the liver, pneumonia, encephalitis, and brain tumors.

Diagnosing patients would never be the same, nor would interviewing them. Patients who expected open-ended psychoanalysis were surprised to find doctors boxed in by literal boxes — doctors were provided diagnostic criteria to tick off one by one, a process that some have called "the Chinese menu" approach. It may not have been creatively fulfilling, but now there were strict boundaries in place that kept psychiatrists from drawing outside the lines if they wanted reimbursement from insurance companies, who had fully embraced the manual. The goal was to make diagnosis standardized in such a way that someone in Maine who was diagnosed with schizophrenia would be diagnosed using the same criteria as someone in Arizona, ensuring that psychiatrists on either side of the country had a far greater chance of making the same diagnosis if they were faced with the same patient. Doctors now had a shared language. Reliability.

Like it or not, this is what a revolution looks like.

"It is as important to psychiatrists as the Constitution is to the US government or the Bible is to Christians," wrote psychotherapist Gary Greenberg. All drug trials from the birth of the *DSM-III* forward were based on the manual's criteria; insurance companies used it to decide how much coverage a person should receive; if a shrink or any kind of mental health professional wanted to get reimbursed for their time, they'd better know how to cite the *DSM* from memory. The *DSM-III* turned madness into different types of disorders that each responded to specific drug treatments, creating "rich pickings for the pharmaceutical industry." And it didn't stop with psychiatrists, extending to psychologists, social workers, and lawyers. It's used in everything from criminal cases to custody battles, from courtrooms to the allocation of special needs resources in public schools.

One of Spitzer's pet projects was to define *mental disorder,* a pursuit that he had been fixated on since the homosexuality debacle. The *DSM-III* laid that out at the very outset: A mental disorder "is conceptualized as a clinically significant behavioral or psychological syndrome or pattern that occurs in an individual and that is typically associated with either a painful symptom (distress) or impairment in one or more important areas of functioning (disability)." Not only did it associate mental illness with dysfunction, which was meant to protect us against making illnesses out of healthy eccentricity, but it also located the cause of mental illness inside the person (not with overbearing mothers or weak fathers, for example) in the same way that physical diseases, like cancer or heart disease, affect the body. So the manual used the term *disorder* — which implied a stronger biological connection — and threw away *reaction,* a relic of the psychodynamic era.

The *DSM* said outright that the continued distinction between physical and mental, between organic and functional, was "based on the tradition of separating these disorders," while acknowledging that these distinctions were somewhat arbitrary. "Hence, this manual uses the term 'physical disorder,' recognizing that the boundaries for these two classes of disorders ('mental' and 'physical' disorders) change as our understanding of the pathophysiology of these disorders increases."

To reflect this, the manual did not provide causes for the psychiatric disorders listed — the science just wasn't there. The goal instead was to keep that part open-ended until the science caught up. It's unclear if the clinicians who bought these books took note of these caveats, however, because everyone else saw the manual, combined with the promise of emerging neuroscience and genetics, as a recasting of psychoanalytically interpreted illnesses into full-blown brain illnesses.

No matter how little proof was there, psychiatry fully embraced the illness model — also known as the field's remedicalization. Harvard psychiatrist Gerald Klerman called it "a victory" for science. It altered the way both doctors and patients saw the provenance of illness and their roles in it — instead of repressed egos and ids or frigid mothers, you had screwed-up brain chemicals or faulty (but not our fault) wiring. Psychiatrists like Nancy Andreasen saw this as a step forward for patients who "no longer must carry the burden of blame and guilt because they have become ill." And that the world should "behave towards a patient just as they would if he had cancer or heart disease."

All the while, the problem of Rosenhan and his pseudopatients nettled the manual's creator. As Spitzer worked on drafts of the *DSM,* he often returned to Rosenhan's study and asked himself: *Would David Rosenhan and his pseudopatients get past this one?*

"When we would write a criterion, for instance, we would often have the study in the back of our minds," explained Spitzer's wife, Janet Williams, who also worked on the *DSM-III.* "Criterionating, we used to call it. You had to write the criteria down and then think of every which way to question it, to improve it . . . We were always asking those things. This was when Rosenhan would inevitably come up."

Spitzer was determined to make sure that the publicity nightmare that Rosenhan and his seven pseudopatients generated would never happen again. "Rosenhan's pseudopatients would never have been diagnosed as schizophrenic if the interviewing psychiatrists had been using *DSM-III,*" wrote Tanya

Marie Luhrmann.

"What Bob [Spitzer] did," psychiatrist Allen Frances said in an interview, "was change the face of psychiatry, change the face of how people saw themselves. It wasn't just a plus, but he did change the world, and that change was very much instigated by the Rosenhan project." Without Rosenhan's study, Frances told me, "Spitzer could never have done what he did with the *DSM-III.*"

It seemed to be a win for us all. Now we had a solid diagnostic system; we had medical language that replaced psychobabble; we had reliability so that doctors all over the world would make a consistent diagnosis.

It sounded, at least at first, like progress to me. I've met some of the holdover psychiatrists from the psychoanalytic era — one told me that he used to get an erection while standing at a podium in front of a new class of medical students and that he'd show it off by jutting out his hips and walking up and down the aisles. Another told me that I was fully healed from autoimmune encephalitis not because of advances in immunology or cutting-edge neuroscience, but because I "hadn't experienced any real trauma before that moment." As if a five-minute interaction can reveal something so deeply rooted.

If this arrogance is what the *DSM-III* replaced, good riddance.

21
THE SCID

In 2016 Spitzer's wife, Janet, invited me to attend his memorial lecture at the New York State Psychiatric Institute, his long-term employer. On my way to the lecture, while wandering along a cul-de-sac formed by a group of identical academic buildings, I lost my way and asked two young men, who looked like medical interns or residents, where I could find the institute. They pointed me to a building at the end of the street and waved as I walked off.

Their helpful responses reminded me of Rosenhan's mini-experiment in "On Being Sane in Insane Places." In the experiment's first iteration, research assistants posed as lost students at Stanford Medical School and were catered to with a pushy level of politeness. In the second iteration, Rosenhan had his pseudopatients ask staff for directions and then monitored the responses. Rosenhan included this interaction from his Haverford hospitalization in his published paper:

Pseudopatient: Pardon me, Dr. _____. Could you tell me when I am eligible for grounds privileges?

Physician: "Good morning, Dave. How are you today?" (moves off without waiting for a response.)

(It is worth noting that all I could find in Rosenhan's notes were the students who had conducted the experiment in the medical school — there is, frustratingly, zero conclusive evidence beyond what he wrote in the study that Rosenhan or the other pseudopatients actually conducted this experiment inside the psychiatric hospitals.)

When I finally arrived at the memorial lecture, the auditorium was packed. Dr. Michael First, a close colleague of Spitzer's, opened with an overview of Spitzer's work. Guess who made the cut?

"The following year David Rosenhan published a controversial paper in *Science* describing how eight pseudopatients were admitted to psychiatric wards for an average of nineteen days despite behaving normal after a single initial claim of hearing a voice that said 'thud,' " Dr. First said. On my recording you hear my laugh. Rosenhan had wormed his way into Spitzer's bio. "Now Bob wrote a scathing critique of this study, and this is a quote — and I like this quote because this is a typical way of Bob in his artful way of using language of sort of putting the study down. He said, 'A careful examination of the study's methods, results and conclusions leads me to a diagnosis of 'logic in remission.' "

The room erupted in laughter. It still killed.

Dr. First finished his short introduction and called on Dr. Ken Kendler, a researcher and professor of psychiatry at Virginia Commonwealth University who contributed to the *DSM-III-R* (the revision of *DSM-III*) and *DSM-IV,* and chaired the *DSM-5*'s Scientific Review Committee. (I'm giving you this background because it makes what comes next all the more surprising.) I expected his lecture to be a "rah-rah-rah!" celebration of psychiatry's bible. I was mistaken.

Ken Kendler has the kind of mind that expects you to rise to its level, but for our purposes I'll attempt to sum up. Basically, he told the audience that in the process of legitimizing the *DSM,* psychiatrists took it literally, ignoring all the gray unknowns.

Psychiatrists believed in the "reification of psychiatric diagnoses." Or, in my words, psychiatrists got high on their own supply and started to believe that there was more *there* there. "We were really proud of our criteria when these came out and that kind of added to the sense that we really wanted a glow around these [diagnoses], to say that these are 'real things,' we've really got it here, it's all in the manual," Dr. Kendler said. "Kind of like Moses coming down from Mount Sinai, except it was a Jewish guy called Bob Spitzer."

When Spitzer brought his tablets "down from the Mount" in the form of the *DSM-III,* the field embraced the manual with an almost religious devotion. "We ask people: Are you sad? Are you guilty? Is your appetite down? We're struggling as a field. Symptoms and signs are all we fundamentally have," said Dr. Kendler. Though the symptoms and signs are very real, the underlying causes remain as mysterious as they were a century ago.

The *DSM-III* did fundamentally change mental health care in this country — but many experts now question if the change was the right direction. "Rather than heading off into the brave new world of science, *DSM*-style psychiatry seemed in some ways to be heading out into the desert," wrote Edward Shorter in his *A History of Psychiatry.* "The sheer endlessness of the syndrome parade caused an uneasy feeling that the process might be somehow out of control."

It's easy to forget that all of the major psychiatric diagnoses were designed and created by consensus. Creation was neither smooth nor orderly. A core group of less than ten people, most of whom were psychiatrists, "clustered around Spitzer, all of them talking as he banged out text on his typewriter. There were no computers, and revisions were made by manual cutting and pasting," wrote Hannah Decker in *The Making of the DSM.* Angry disagreements abounded. Feelings were hurt. All the while Spitzer typed away furiously, a demon on his typewriter getting it all down, devoting seventy to eighty hours a week to the project. "There would be these meetings of the so-called experts or advisers, and people would be standing and sitting and moving around," one psychiatrist who worked on the manual told the *New Yorker.* "People would talk on top of each other. But Bob would be too busy typing notes to chair the meeting in an orderly way." Psychologist Theodore Millon, a *DSM-III* task force member, described the scene: "There was very little systematic research, and much of the research that existed was really a hodge-podge — scattered, inconsistent, and ambiguous. I think the majority of us recognized that the amount of good, solid science upon which we were making our decisions was pretty modest."

Even reliability, trumpeted as one of the major wins of the new manual, was oversold. In 1988, 290 psychiatrists evaluated two case studies and were asked to offer a diagnosis based on *DSM* criteria. The researchers, however, had devised a way to test the clinicians' own diagnostic biases: They created multiple patient case studies out of the two set examples by altering two factors: race and gender. Even when presented with identical symptoms, clinicians tended to identify black men as more severely ill than any other group. (This continues to be true today: One 2004 study showed that black men and women were four times more likely to receive a schizophrenia diagnosis than white patients in state hospitals.)

The issue with reliability is that consensus does not necessarily translate to legitimacy. "In days of yore, most physicians might agree that a patient was demonically possessed. They had good reliability, but poor validity," noted Michael Alan Taylor in *Hippocrates Cried.*

Rosenhan never spoke publicly about his thoughts about the *DSM.* Given his private correspondence with Spitzer, I'm sure he suspected that his paper at least shaped parts of the manual. Would he be proud of the wide-reaching effects of his experiment or would he be dispirited by how his study was exploited to push the field's agenda to save itself?

The next edition, the *DSM-IV,* was overseen by Allen Frances in 1994. "It followed dutifully in Spitzer's footsteps, though it included new diagnoses and broadened and weakened the criteria that had to be met for any particular diagnosis to be assigned," according to sociologist Andrew Scull.

As we saw, diagnostic boundaries for mental illness have collapsed and expanded over time. When Rosenhan was hospitalized, the schizophrenia diagnosis cast a far wider net than today. *How shall we know them?* Make that bucket too wide and these words become meaningless; make it too narrow and you miss people who desperately need help. Dr. Keith Conners, considered the "godfather of medication treatment for A.D.H.D." who helped establish standards for diagnosis of the condition, expressed dismay at the growing numbers of kids (15 percent of high schoolers) with the label. "The numbers make it look like an epidemic. Well, it's not. It's preposterous," he told the *New York Times* in 2013. "This is a concoction to justify the giving out of the medication at unprecedented and unjustified levels."

When the fifth edition of the *Diagnostic and Statistical Manual of Mental Disorders* came out in 2013, it crash-landed to terrible press. The *DSM,* behind schedule and belabored by criticisms from inside (and outside) its own ranks, aimed to implement a "dimensional aspect" or a continuum of mental disorders rather than the strict categories that defined the previous volumes. At least three books in 2013 slammed the manual before it was even published — Gary Greenberg's *The Book of Woe,* Michael Alan Taylor's *Hippocrates Cried,* and Allen Frances's *Saving Normal.*

Saving Normal, which Frances described as "part mea culpa, part j'accuse, part cri de coeur," all anti-*DSM-5,*[1] was the most vociferous given his former position as the

1. The American Psychiatric Association dropped the use of the Roman numeral system for *DSM-5* to make it easier to add "piecemeal revisions in

head of the *DSM-IV* task force and his close relationship with *DSM* godfather Spitzer. It was, of all people, Spitzer himself who had recruited Frances from out of retirement to join him in warning the public that the new manual would likely "produce a very dangerous product." The release of the manual was stalled twice — thanks, at least in part, to these two heavyweights. Frances wrote open letters to the APA, op-eds, and tweets. He admitted to the public that he had failed "to predict or prevent three new false epidemics of mental disorder in children — autism, attention deficit, and childhood bipolar disorder." Diagnoses of childhood bipolar disorder had increased fortyfold in the eight years between 1994 and 2002; there had been a fifty-seven-fold increase in children's autism spectrum diagnoses between the 1970s and today; and attention-deficit/hyperactivity disorder, once a rarity, now affected an estimated 8 percent of children between the ages of two and seventeen. Frances's point that our definitions have drastic, real-life implications was a valid one — were we reaching people who had long been ignored or were we overdiag-

---

the future" in software updates, sociologist Andrew Scull explained.

nosing and overmedicating children? Frances warned that the *DSM-5* would further "mislabel normal people" and create "a society of pill poppers" (in a time when already one in six adults were using at least one drug for psychiatric problems). Some APA psychiatrists reacted by arguing that Frances had not only a reputation to save but also money to lose, because the new manual would reduce the royalties he was collecting on his own creation, the earlier version of the book.

Still other greats in the field piled on. Dr. Steven Hyman, director of the Stanley Center for Psychiatric Research at the Broad Institute at MIT and Harvard, called it "an absolute scientific nightmare." Dr. Thomas Insel, the former director of the National Institute of Mental Health, said that the manual had a "lack of validity" and was "at best a dictionary." Here's the deal: The science wasn't there when Spitzer and company wrote the manual (and they tried to acknowledge this by leaving the manual open for revisions). Despite all the effort in the three decades since, it still isn't there.

Many research psychiatrists I've interviewed liken *DSM* diagnoses to our understanding of headaches — we have symptoms with no knowledge of the underlying cause.

You can, for example, think you have merely a headache when you in fact have a brain tumor. Pop an Advil and your headache might go away, but you've still got a metastasizing mass in your skull. Without a way to find that tumor, how do we tell the difference?

The most concerning part, from my perspective, is that the *DSM* approach rendered the practice so rigid, so fixed, that the patient, the person, the human, was lost. As I would learn, this doesn't just affect the relationship between doctor and patient, but can increase misdiagnosis.

I had tested this out myself with Dr. Michael First, the man who introduced Spitzer and mentioned Rosenhan at the memorial lecture.

"I'm nervous," I said as I turned on the tape recorder in First's office. "Why am I nervous? Have you been SCIDed yourself?"

"Nope," Dr. First said.

Dr. First is not exactly warm and fuzzy — he's hyperclinical and a straight shooter, two things that have made him key in the creation of the last three incarnations of the *DSM* — but the chunky metal ring I spotted on his finger during our interview betrays what I interpret as his softer,

Woodstock-hippie vibe. He is often called upon to consult in high-profile criminal cases, recently that of the murder of six-year-old Etan Patz, which ended in a hung jury (the defendant was found guilty in a second trial). But his main contribution to the *DSM* world is the SCID — the Structured Clinical Interview for *DSM* — a pre-written set of interview questions designed to make a psychiatric diagnosis based on *DSM* criteria. I had asked if he would be open to SCIDing me about my experience with psychosis, pretending that he didn't know the diagnosis. Dr. First seemed open to a challenge — even if the odds were stacked against him.

In 2008 he appeared on a BBC reality show called *How Mad Are You?* where ten people — five "normal" and five who had been diagnosed with psychiatric conditions — lived in a house observed by a psychiatrist (Michael First), a psychologist, and a psychiatric nurse, and engaged in a variety of tasks, including performing stand-up comedy and cleaning out cow stalls. The panelists' goal was to ferret out the mentally ill and correctly label them with only five days of observation. The panelists didn't do such a hot job. They nailed the guy with obsessive-compulsive disorder after watching him struggle to clean up cow manure, but incorrectly diagnosed one volunteer with bipolar disorder (where no disorder existed) and one with a history of schizophrenia (there was no history). It's worth recognizing how astonishingly deep Rosenhan's thesis has cut: Despite all of psychiatry's efforts to legitimize itself in the time since, the impossibility of distinguishing sanity from insanity had received the most mainstream of honors — its own reality show.

First began. "Okay, I'm going to do it straight through as if we're doing it for real because we are doing it for real."

He rattled off the first few questions and I answered them in quick succession: "How old are you?" "With whom do you live?" "How long have you been married?" "Where do you work?"

I explained that I had been dating my husband for seven years, but that we had met when I was seventeen. I told him about our recent marriage. He asked about work so I summarized my history at the *New York Post,* where I had worked for even longer than I'd known my husband.

"Have you ever had a period of time when you couldn't work or go to school?"

"Yes," I said. "When I was sick."

"Tell me about the illness."

I walked him through the natural history of my illness, starting with the funk of depression, which morphed into mania, then to psychosis, and finally to catatonia before I was accurately diagnosed with autoimmune encephalitis. He asked questions along the way but kept me on the straightest road possible. He maintained an emotional distance — never a *wow* or *that must have been hard* or even a *how did that make you feel?,* all common reactions among others hearing this story. He moved right along, question after prewritten question.

"Have you ever wished you were dead or would go to sleep and never wake up?" he asked.

I thought of Rosenhan's answer to this question posed during his intake interview at Haverford Hospital. I responded no.

"Have you ever tried to kill yourself? Have you ever done anything to hurt yourself?"

No, no.

"Have you had any problems in the past month?"

"Problems?" I asked.

"Anything, like at work, home, other problems."

"Every day I have a problem." I laughed.

What kind of question was this?

"Like everyday normal stuff?"

"Yes."

"How has your mood been in the past month?"

"Actually pretty good," I said. "I've been meditating."

"Medicating?"

"No, meditating."

Moving on.

An odd dynamic was occurring — I said no to all the above, but despite myself I found that I wanted to please this doctor. I didn't want to disappoint him with my normalcy.

"In the past month, since the twentieth of March, was there a period of time when you felt depressed or down for most of the day nearly every day?" This was odd. I had just told him that my mood had been good thanks to that Headspace app. He was just reading off the page.

"In the past month, since the twentieth of March, have you lost interest or pleasure in things you usually enjoyed?"

It now felt like what I imagined a courtroom interrogation would be like — as if he was trying to catch me in a lie.

He went on to ask the same questions but extended over the course of my life. During my illness, for example, I felt depressed but a yes wasn't enough. He wanted to know exactly how long I felt blue, as if emotions have hard edges.

"A week, that's it?"

"Oh, I don't know. Maybe a month? It's so hard for me to say."

"In the hospital were you depressed?"

"I was so cognitively impaired. People said I was, but I don't remember."

"How about the mania?" he continued. "How long did that last?"

"Again, it was so mixed with depression it's hard to say." I'm trying desperately to make something concrete out of something that just isn't. Emotions are not mathematical formulas, inserted as x + y = psychiatric diagnosis.

"Just to recap. February 2009, three weeks most of the day every day you were depressed. Does that sound right?"

"Sure."

He focused on the first two weeks of the depression and I played along, as if I, or frankly anyone else, could accurately respond to such prescribed questions about such an irrational, frightening time.

"How long did the mania last total: a week and a half?"

"It's so hard to say . . ."

"During that week-and-a-half period, how did you feel about yourself? More self-confident than usual?"

"Sometimes. But one second I'd be the best and then the worst."

"But certainly for a significant amount of that time you did have that feeling."

"Sure." It was astounding. Everything needed to be so concrete.

More questions: "Sleep? Concentration? Spending more time thinking about sex? Pacing? Buying things you couldn't afford?" And then my favorite one: "Did you make any risky or impulsive business decisions?" This after telling him that at the time I made thirty-eight thousand dollars a year. I laughed at this one. "Oh, all those risky business investments!"

"Now I'm going to ask you about some unusual experiences," he said, again reading. "During that time, did it seem like people were talking about you?"

"Yes. Nurses were talking about me. I could read their mind."

"Did you have the feeling that some things on the radio or TV were especially for you?"

"Yes," I said. "I had a whole delusion about the television and my father."

"What about anyone going out of their

way to give you a hard time or trying to hurt you?"

The yeses kept coming: "Did you ever feel like you were especially important? Had special powers?"

Of course: I vividly remembered my brief brush with godlike powers when I believed I could age people with my mind.

"Were you ever convinced that something was wrong with your physical health even though doctors told you there wasn't?"

My obsession with bedbugs; my conviction that I was dying of melanoma.

"Were you ever convinced that your boyfriend was being unfaithful?"

The time I rummaged through his things in search of nonexistent clues to his imaginary affair.

There were specific questions about people implanting thoughts in my head, about the porousness of human interactions, about unrequited love, that didn't fit. At the end of our interview, Dr. First closed the book.

"If I didn't have the answer" — meaning autoimmune encephalitis — "I would have a different diagnosis. This would have been schizophreniform disorder."

Schizophreniform is when someone exhibits features of schizophrenia for less than six months, the minimum length of time required for a schizophrenia diagnosis. (Though this minimum time length was created under the Feighner criteria, which predates the creation of the *DSM-III,* I suspect it was included in the *DSM* at least in part thanks to Rosenhan's study. If you needed to exhibit symptoms for six months, then the pseudopatients, who were supposed to only very recently have started to hear voices, would have at least been filtered through a less definitive diagnosis.)

When I told him that the psychiatrists at the hospital offered two diagnoses, bipolar 1 and schizoaffective disorder, he reopened his book. "If you were depressed at the same time you were psychotic . . . That would make sense . . . Actually, it wouldn't have been schizoaffective because the amount of mood wasn't as long as the psychosis. Was there a time when you were psychotic when your mood was normal?"

I laughed here. "Can you be psychotic and your mood be normal? Is that even possible?"

"Well, yes," he said. "I think technically it really wouldn't be schizoaffective. Technically it's kind of mixed. It's hard to say. That's the problem. You really need to know with a reasonable amount of precision . . ."

I couldn't believe it. I had a more precise

view of my illness than most — especially a psychiatric one — since I had spent a year writing and researching it and the past four years talking endlessly about it. I still couldn't adequately answer his rigid questions.

"At the time, the two diagnoses that would have been most reasonable were schizophreniform and schizoaffective disorder," he said. "But it doesn't matter because both of those diagnoses are wrong." He closed the booklet. It was brave and honorable to be so candid about the limitations of his creation. He continued. "We see this all the time with people with psychotic symptoms that don't respond to antipsychotics. Is it because they really have your condition? Or that some people with bona fide schizophrenia don't respond? Or maybe what we're calling schizophrenia is actually many different things."

He had dropped the formality of the interview, to my relief. "You can see how messed up this field is," he said.

A moment of awkwardness passed before I removed my wallet. "So how much do I owe you?"

"Well, my typical price for this kind of thing is $550."

Five hundred and fifty bucks for him to

give me a misdiagnosis. I couldn't believe it. And I don't think he could, either.

"Do you take Amex?"

■ ■ ■ ■

PART FIVE

■ ■ ■ ■

The greatest obstacle to discovery is not ignorance — it is the illusion of knowledge.
— Daniel Boorstin

22
THE FOOTNOTE

The harder I fought to make sense of it, the more I realized that Rosenhan and his study were like quicksand: Whenever you felt that you were on solid ground, the support would fall away, leaving you deep in the dark muck and sinking fast.

Thanks to Bill Underwood, I learned the first name of a fellow graduate student who also took part in the study: Harry. I scanned the 1973 Stanford psychology graduate student class and there he was, just a few spots above Bill: Harry Lando. I noticed immediately, however, that Harry's name didn't match the first name of any of six remaining unidentified pseudonyms — he wasn't John or Bob or Carl — and his position as a graduate student didn't match their bios, either. Was I wrong that the first names were kept the same? I searched "Harry Lando" on PubMed and found around a hundred studies on smoking cessation but nothing on Rosenhan. On World-Cat, I typed in "Lando" and found more smoking articles, but then I added "Rosenhan" and bingo — a hit for a study titled "On Being Sane in Insane Places: A Supplemental Report," published in *Professional Psychology* in February 1976. The summary read:

The author gives the psychiatric institution a favorable review after spending 19 days as a pseudopatient in the psychiatric ward of a large public hospital. He recommends stressing the positive aspects of existing institutions in future research.

There he was. Another pseudopatient: *He recommends stressing the positive aspects of existing institutions in future research.* Out of the 1,066 results for Rosenhan on World-Cat, Harry Lando's study was number 251 on page twenty-six. I had skimmed past it in my initial search, long before I'd begun looking for pseudopatients, and in all the digging since had not encountered even one source that had quoted it.

I tracked down a hard copy of the study, which featured a black-and-white author photo of a young man with a thick head of hair, a big bushy mustache, and an angular

374

face, and read the opening sentence: "I was the ninth pseudopatient in the Rosenhan study; and my data were not included in the original report."

Of course! The footnote! "Data from a ninth pseudopatient are not incorporated in the report because, although his sanity went undetected, he falsified aspects of his personal history, including his marital status and parental relationships. His experimental behaviors therefore were not identical to those of the other pseudopatients." Harry Lando didn't match any of the eight pseudopatients because he *wasn't* one of them. He was the unknown ninth — the footnote who received little attention in rehashings of the study because it was used as pro forma acknowledgment that the data was so pristine, Rosenhan had thrown out a whole data set that didn't uphold the study's standards.

However, knowing what I now knew, that logic sounded a bit sanctimonious — Rosenhan himself had done that exact thing, misleading doctors about his own symptoms and changing his medical records.

Even more intriguing than the hypocrisy was the question: Why would Harry be advocating *for* the institutions instead of railing against them? He used words like "excellent facility" and "benign environment," a drastic departure from the experiences of Rosenhan and Bill, the other two patients I'd found so far.

I located Harry's picture, his face a bit older and without the bushy mustache, on the School of Public Health's website at the University of Minnesota, where he now works as a psychology professor with a focus on the epidemiology of smoking behaviors. I sent Harry an email. Three days later I found myself ear-to-ear with the second of Rosenhan's mystery pseudopatients. By this point my level of enthusiasm, which, let's be honest, is normally set somewhere around eight, was probably measuring on the Richter scale. If it's possible to hear someone beam, that is how you would describe my rambling, rapid speech. We discussed Bill, whom Harry seemed delighted to know that I had already contacted; my own experience with autoimmune encephalitis, which he seemed interested in; and then we got down to business.

Harry's experience was, indeed, incongruous with Bill's. He was also a very different guy. He's the kind of person you might call an absentminded professor. One of his regrets as a kid, he said, was that he was not enough of a rebel, too much of a goody-goody.

Harry's career studying the mind was born from the most universal of urges: He had developed a crush on a junior professor who suggested that he take graduate-level courses while an undergraduate at George Washington University in Washington, DC. One of these high-level courses was taught by Dr. Thelma Hunt, the youngest person to be awarded a PhD at the university during a time when women didn't often receive such honors at any age. Though her accomplishments were legion during her fifty-nine-year career (including establishing therapy programs and recruiting more women to the sciences), one of her most cited works was with Walter Freeman, Rosemary Kennedy's doctor, the pioneer of the transorbital "ice-pick" lobotomy. They collaborated on *Psychosurgery: Intelligence, Emotion and Social Behavior Following Prefrontal Lobotomy for Mental Disorders,* which featured three hundred pages of case studies and photographs of post-lobotomy patients. Hunt supplied supplemental materials on cognitive and intelligence studies performed after the surgery, measuring a patient's "self-regarding span," or the amount of time a patient would talk about herself, pre- and post-operation. Patients before the surgery spoke about themselves for nine minutes, on average; post-op this dropped to four minutes for a standard lobotomy, two minutes for radical ones. I'm not sure what this tells us about what lobotomy does to the self — but it's safe to say it's not good.

Harry didn't remember much about Dr. Hunt's class other than it was so dry that it put quite a few students to sleep (though it wasn't boring enough to deter Harry from pursuing a higher degree in psychology). He applied to a PhD program at Stanford to study social learning theory with psychologist Al Bandura, well known for his "Bobo doll study" of aggressive behavior in preschool children. (Among the study's findings was that when preschoolers at Stanford's Bing Nursery School watched adults physically or verbally abuse a three-foot blow-up cartoon clown, they would mimic the assaults, an example of behavior modeling, showing that abusers are often made in early childhood. In some ways, it was in line with the question that many postwar social psychologists, like Milgram with his shock machine and Zimbardo with his prison studies, were pursuing: Are you born bad or are you made that way?)

Despite his research interests, Harry didn't acclimate to Stanford the way Bill had. Harry, who was a few years into an unhappy marriage, found Stanford to be an unfriendly, stifling, overly competitive place. Like Bill and Maryon, he joined a few sit-ins protesting the Cambodian incursion and, later, a mass demonstration honoring the victims of the Kent State shooting, but mostly he felt lost. "I was, I would say, quite insecure. I wondered if I really belonged at Stanford," he told me. "I wondered if they would discover my incompetence." When I asked him if he maybe was depressed, he had to think it over. "I don't think I would have met the criteria for clinical depression," he said in his detached manner, "but certainly I was not a happy camper."

He even found the work unfulfilling. Bandura, though renowned for his Bobo doll experiments, was studying aversion therapy when Harry joined his research team. Harry soon learned that prepping participants for experiments that would likely torture them didn't stoke his enthusiasm. He nearly quit when he had to clean up the mess made by one of the subjects, who, after inhaling dozens of cigarettes in a row for a study on smoking aversion, deposited the contents of his stomach into Al Bandura's snake cage. Harry certainly hadn't dreamed of a career cleaning up vomit; he wanted to tap into something greater than himself. In his free time, he read *One Flew Over the Cuckoo's Nest* and *I Never Promised You a Rose Garden,* two books in heavy rotation on the campus at the time, along with the usual suspects like Goffman's *Asylums,* Laing, Szasz, and Foucault.

In the fall of 1970, Harry enrolled in a graduate seminar called Psychopathology. He didn't recall many specifics from Rosenhan's lectures except the awe he inspired. At one point, Rosenhan invited the eight students in the class over to his house. Mollie's food was in top form that night — Greek egg soup, lemony and creamy — and while the students devoured it, Rosenhan made his pitch. Harry was so blown away by the food, by the house, by the Meyer lemon and pomegranate trees, by the black-bottomed pool in the backyard, by Rosenhan, that, he said, he would have signed on any dotted line: "It was kind of like, *Wow, this is exciting.*" Rosenhan had set his hooks in him and he was intrigued by the thought of going undercover. It's not in Harry's nature to overanalyze, but it seems clear that he joined the study because Rosenhan gave him an opportunity to belong to something. Thanks to scheduling conflicts and other halfhearted excuses, however, none of the other students ended up going through with hospitalization except Bill Underwood.

Harry chose the name "Harry Jacobs," which he said just popped into his head, and Rosenhan and his research assistant gave him a fake address that was near their target hospital, Langley Porter, the University of California at San Francisco's psychiatric hospital, the oldest one in California. Like Bill, Harry did not recall that Rosenhan prepped them beyond a tutorial on how to cheek pills. "That kind of surprised me. I felt like we had pretty minimal coaching. I met with Rosenhan the day before. He said something in class about 'empty, thud, and hollow,' the voices, existential psychosis, but really I got maybe fifteen minutes of coaching, and it really made me nervous because, you know . . . having not been a clinical psychology student, having grown up with the ideas that people who are mentally ill are just really off the wall, right? So I almost felt like I was going into the lion's den here. I mean, what would the patients be like?"

Harry had remembered a PSA from his childhood, which warned children that they should be kind to people with mental illness because *it could happen to you.* These ads scared the daylights out of little Harry, who developed a phobia that he would one day be boarded up in an insane asylum and "catch" mental illness. Now here he was nearly two decades later volunteering to enter one.

It was late November, after Thanksgiving, a perfect autumn morning in San Francisco. Harry put on his slacks and a dress shirt (he was not of the hippie clique, didn't have a mustache, hadn't grown out his hair). He carried very little money, enough to get him to the hospital and a little extra pocket change, and no identification just in case they searched him.

He took a bus to the admissions office in San Francisco's Langley Porter. A nurse there asked if he had an appointment. He replied that he didn't, but that his psychologist, Dr. David Rosenhan, had referred him. When Harry told them his address, the nurse replied that he would have to go to San Francisco General Hospital because Langley Porter was not in his (fake) address's catchment area. The nurse gave him bus route directions and sent him on his way. Harry left the hospital and went to a phone booth to call San Francisco General.

The operator debated if they would take him given his address, took his number, and told him that someone would call him back. Harry, now thoroughly unnerved, phoned Rosenhan's research assistant, the study's point of contact, a young woman whose name escaped him, and delivered the bad news. She sounded disappointed but told him to hang tight.

Seconds later the phone rang. Harry was rattled when a strange voice introduced himself as a psychiatrist from yet another hospital. How did this psychiatrist get the phone booth number? Who had contacted him? What hospital was he calling from? And what exactly transpired in this phone call? Harry couldn't remember, though he was sure he delivered his practiced script — *thud, empty, hollow.* Still, something about his story or the way he was telling it made the doctor think that Harry was a suicide risk. I pressed Harry repeatedly to explain why the doctor might feel that way, but Harry could not come up with any reason for it.

"You're forcing my hand," the doctor kept repeating, as Harry recalled. "You're forcing my hand. You need to come to the hospital."

Harry boarded another bus and tried to

tamp down the growing unease. This was not a vetted hospital. He had no clue about how this hospital worked or what kind of patients it treated.

He was heading straight into the great unknown.

Harry didn't remember walking in. Somehow he ended up on the fourth floor in a private room, an office, where a psychiatrist sat behind a big desk with some personal touches, a few family photos, a book or two. The psychiatrist asked Harry to sit in the seat opposite him. Harry felt sweat staining his undershirt, but the funny thing was, he didn't feel nervous. It was as if he was observing the anxiety from a great distance. It felt like a schoolyard game, like he was up to bat, about to take a swing for the fences. The words flowed: He was Harry Jacobs, a Berkeley grad (he changed his school from Stanford to his wife's school), who had begun to hear voices just a few weeks before, that said, "It's empty. Thud. It's all hollow." All parts of the approved script.

There were amendments, Harry admitted. He told the psychiatrist that he lived alone off campus, leaving out the fact that he lived with *his wife* off campus — which,

Rosenhan would write, gave evidence of isolation that could spell out trouble for someone presenting with a severe psychiatric symptom (even though Bill had done exactly the same thing and left Maryon out of the narrative). Harry then told a more significant lie. He said that he had no close family since his parents were killed in a car accident a year before (in reality his parents were alive and well). Why did he do this? Even Harry didn't have an answer, but he did insist that Rosenhan approved the changes to his biography prior to going to the hospital. Rosenhan's notes (he used the pseudonym "Walter" for Harry in some scrap notes, and because he did not indicate that Walter was the footnote, I had assumed at the time that Rosenhan had used another pseudonym for one of his original eight) tell another story: "Just why Walter changed his script was never clear, but I strongly suspect that it was because he wanted very much to be hospitalized, and like the rest of us, expected that he would not be admitted on the basis of such a slender set of symptoms . . . These alterations of script made it impossible for me to include his data in the study, since I could not know what impact the changes had on the staff's perceptions." And yet, again, we know too well the distortions Rosenhan made to his own script.

Either way, once admitted, Harry became an interesting case. About fifteen minutes into his intake interview, the psychiatrist asked if he could bring in two other psychiatrists to consult. Harry felt flattered by the doctor's interest. When they asked about how he spent his days, Harry answered honestly: the bleakness of holing himself up in his apartment watching television, the endless studying, the hypercompetitive atmosphere at school, the absence of close friends. Harry spoke about his feelings of worthlessness and self-doubt. It wasn't until this moment when he was pretending to be Harry Jacobs in front of three psychiatrists that he realized how truly miserable he was. "I was not a very happy graduate student, and at the time being unhappily married didn't really help the situation. But part of it . . . is a lot of self-doubt, being with these famous people in the department, you know . . . There was this feeling of isolation." He may not have lost his parents or lived alone, but the heartsickness was genuine; only the cause was fabricated. Was this really worse than Rosenhan's own insistence that he put copper over his ears to drown out the noises?

After forty-five minutes, Harry was admit-

ted to U.S. Public Health Service Hospital.[1]
"I felt like I passed a test."

The word that best describes Harry's first impression of the ward is *light*. A bank of windows in the dayroom let in cascades of natural illumination that made the ward seem impossibly uplifting. Christmas decorations, wreaths and handmade ornaments, and a tree strung with lights gave the ward a festive, joyous feel. Was this really the horror house he'd imagined in his youth?

Men and women shared the unlocked floor and were free to roam around as they pleased. The nurse took him on a tour (this in itself was unusual; it didn't happen with Rosenhan or Bill) and explained that there were wake-up and sleep times, but what you did in between was up to you. And there were no uniforms! The staff wore street clothes. Harry got them mixed up with

1. Just before this book went into production and while he was preparing to move, Harry chanced upon the notes he took during his hospitalization. These notes confirmed (after much debate) which hospital Harry visited: the U.S. Public Health Service Hospital in northwest San Francisco, a federally funded research hospital that originally catered to soldiers and officers in the Navy.

patients more than once, especially in the beginning. Perhaps the differences could be chalked up to the fact that Harry's hospital was an acute psychiatric facility, intended to provide short-term care with a focus on releasing patients to their homes, outpatient facilities, or, if necessary, state hospitals. This was neither a place of last resort nor a custodial care center into which people would disappear for months or years. The hospital wanted you in and out, and in the process tried to make your experience as pleasant as possible.

In our conversation, Harry didn't remember much about that first night. Rosenhan wrote that Harry's first meal was "eaten in total silence," a reflection of his early uneasiness, though Harry said that his silence might have been a result of the surprise at how tasty his fillet of beef was. His medical file acknowledged this nervous energy: "He engages in finger-cracking," according to Rosenhan's notes.

Like Rosenhan, Harry spent the early days avoiding his fellow patients. Once he started attending group therapy sessions, however, he had no choice but to interact. The patients were mostly around his age, a few younger and a few his parents' age, but none older than that. Some fit the stereotype of a Bay Area hippie circa 1970. Some ended up there after a suicide attempt — newspapers at the time frequently reported people being talked down from the Golden Gate Bridge. One kid, a former member of the Coast Guard, spent eight months on a tiny island in the Pacific, flipped out, and ended up there, guitar in hand. Harry felt a special fondness for that cracked guardsman, who reminded Harry of his brother. Once he plugged into the floor's energy, he realized that he was part of a love-in — people sat in groups, singing, crying, laughing, all part of a community of people who had lived through some heavy stuff.

Pretty much everyone on the ward was against the war, even the nurses. When news of war casualties aired on TV, a nurse commented: "I'm moving to the North Pole," which cracked everyone up. Nearly everyone, anyway. John, a Korean War vet, took an instant dislike to Harry and harangued the whole group for their antiwar views, repeating: "Anybody who is against the war should be shot."

This didn't scare Harry; John seemed like a curmudgeon, not the kind of mental patient you saw ranting in the movies. Harry recalled that the "craziest" person on the ward was a suicide risk named Ray. He was the only patient who wore a hospital gown intended to prevent escape. Before his hospitalization, he had jumped out of a fourth-floor window and survived. He had a few broken bones to prove it. Still, Harry found him to be a pretty rational guy, if a touch blue.

If Ray was blue, Harry was red. Harry described himself as positively amped up those first few days. He had a fire lit inside him that had been absent since he'd moved to Stanford. He wrote nonstop, filling pages in his notebook. (Up until he found them days before this book went into production, Harry was convinced that he had thrown his notes out in a rabid spring cleaning years before I contacted him. Harry does not recall if Rosenhan received a copy of these pages.) The staff noticed. Several approached Harry about his writing and asked him if he was a writer.

Harry presented with enough psychotic symptoms for doctors to prescribe him daily doses of Thorazine. Trouble was, the drugs were not in pill form, but liquid. Liquid Thorazine was introduced in the 1960s as a response to the pervasive problem of patients cheeking pills. The ad campaign in the 1960s read: "Warning! Mental Patients are Notorious DRUG EVADERS."

Harry thought, *Okay, David, what do I do now?* and hesitated for just a moment before he swallowed the unpleasant syrup, grimacing as it slid down his throat, bracing himself for the drug to take effect. Hours later, nothing had happened. "I think that tells you something about my mental state," he said now. Either the dose was so low that it hardly had an effect or he was so unsettled by the environment that the antipsychotic soothed him. Later the doctors switched him to pills that he could cheek so he wouldn't have to test this assumption out.

Harry spent the early days observing, asking questions, but rarely speaking — a behavior that prompted one of the younger, more attractive nurses (on whom Harry had developed a crush) to push him to share more of himself, suggesting that the sublimation of his own feelings was a sign of his suffering. This was an astute observation. He *did* detach — especially at home with his wife. "This really touched me," Harry said.

To Harry's eye, the staff just seemed as though they really enjoyed their jobs. They spoke to patients as equals, engaged them in games and gossip, and even joined their sing-along groups, the sounds of Peter, Paul and Mary filling the ward. When one young female patient was released without a place to call home or any money, one of the nurses took her in until she got on her feet. "The hospital seemed to have a calming effect. Someone might come in agitated and then fairly quickly they would tend to calm down. It was a benign environment," Harry said.

But Harry was still a patient — and he was reminded of this distinction during a meeting with the ward's clinical psychologist, who asked him to draw stick figures, which Harry recognized from the "Draw a Figure" test, a popular psychological test originally designed for children. Harry studied it in graduate school as a tool to assess perception and cognition. Drawing was not one of Harry's talents, and he felt self-conscious. Even though this was make-believe, he still wanted to impress the psychologist — the same way I had wanted to impress First while taking the SCID — and he tried hard to hide his spatial limitation. "I wanted to do the best I could, just as much as if I had been in a 'real situation,' " he told me.

Eventually he asked the psychologist: "Should I continue trying or should I give up?"

The psychologist responded, "It's up to

you." He remembered that response was exactly what he'd been trained to say when faced with that question from a patient. "Having that thrown back at me wasn't entirely pleasant," he admitted.

Early in his stay, a nurse handed Harry his own medical file — an unusual moment in any hospital, let alone a psychiatric one — and told him to walk it to another floor to get an EEG. The minute the CONFIDENTIAL file hit Harry's hands, he knew it was as good as gold. Harry thumbed through the files as he walked. Time was of the essence. They would notice his absence from the appointment if he took too long, but he needed to get this information to Rosenhan. How? A phone! He paced the hallways in search of one, ducking into an empty office, his hands shaking as he picked up the receiver and dialed Rosenhan's number. He didn't remember speaking to Rosenhan, but believed he made contact with Rosenhan's pretty research assistant.

The file confirmed that he was taking antipsychotic medication. Another line read: "Unfit for military service." He couldn't help but think, *Man, this could be helpful.* But then three words jumped out at him: "Chronic, undifferentiated schizophrenia." Rationally he knew that to be hospitalized

he was likely diagnosed with *something* —
but to see it written out in black-and-white
still stunned him.

A new woman joined the group therapy session the next day and turned her back to
the room, refusing to talk. The other patients
devoted the session to cajoling her into
engaging. "We wish you'd join us," they'd
say. Eventually their kindness broke her
down and she began to communicate with
the group, telling them that God had
damned her. One of the patients quoted
Bible passages to her that expressed God's
love and forgiveness. "It's hard to convey
the sense of the beingness of the environment, and how the patients, I mean, how
the patients supported each other. I mean
they cared," Harry said. "I'm getting emotional just thinking about this . . . What
struck me is just how human and I guess
vulnerable the patients seemed."

While Rosenhan had the experience of
wanting to expose himself as a "sane person" ("I am Professor Rosenhan!"), Harry
wanted to confess for entirely different
reasons: "I felt this guilt that they were making such an effort to help me with my
problems, and they were taking time with
me that could have been spent otherwise. I

felt this guilt that I was in the hospital when I didn't need to be. And these were good people . . . I wanted to confess my sins."

A little less than a week in, the ward arranged for a day trip out of the hospital to the beach. The group boarded a shuttle bus and headed forty minutes down the coast. The sea air must have smelled magical, potent with possibility, as they disembarked and made their way onto the beach to enjoy the warm early-December afternoon. Did people whisper, *They're from the loony bin*? If they did, Harry didn't notice. He was too happy. He sunbathed and chatted. It was so much more fun than grading papers in graduate school. All that seemed so far off now. A female patient grasped his hand and whispered: *"Let's stay here. Let's not go back."*

"I felt less de-individualized, more actualized in the hospital with those patients than I did as a graduate student, honestly," he said.

On Rosenhan's notes about Harry's stay, the professor scrawled "HE LIKES IT" on the side of the paper, as if he couldn't imagine such a thing.

By the second week in, Harry had transformed from a shy loner into a ward leader.

His peers seemed to respect him. They looked at him for approval and for advice. He leaned into this newfound position of authority and even dropped little hints that he knew a bit more about psychology than he had let on, offering to administer ad hoc psychotherapy to his fellow patients. Rosenhan interpreted this as trying to differentiate himself from the group. Harry agreed — but saw it in a more aspirational way. "I, of course, imagined myself as McMurphy," he said, channeling the hero from *One Flew Over the Cuckoo's Nest.* "I just got the feeling that patients were looking up to me, and that meant a lot . . . I felt like I could be a positive influence and support for other patients."

He also openly flirted with the young nurse who had coaxed him out of his shell in those early days. "It's hard to concentrate on therapy when you're wearing a skirt like that," he said of her mini-skirt.

She laughed it off, as if they were in a bar, not a psych hospital. Sometimes she would invite him into the nurses' office to relax. John, the military vet who had expressed anger about the war protesters, didn't like the preferential treatment Harry seemed to receive and, one night when he'd had a bit too much to drink on a day-leave from the hospital, said as much. Deep into his cups, John belligerently motioned to Harry inside the station.

"Get out!" John ordered.

"I'm not going to do that," Harry responded, surprised by the power of his new voice. John did not scare him. He was sad, sick, and jealous. (When he later recounted the story, Rosenhan was horrified. "Didn't your dad ever teach you, never confront a drunk?" But Harry had read the situation correctly. John was grumpy but not violent.) As John walked away, Harry relished his newfound confidence. He was changing here — in a positive way.

Two weeks or so in, Harry decided he needed a break. Though he had adapted, he was drained of mental and physical energy. Even in the middle of the night, he was still pretending to be a sleeping patient, and it messed with his head. He decided to push for an early release. As he expected, most of the patients supported his overnight pass. (On this ward, patients helped decide who should receive day and night passes, which contributed to the communal, supportive environment.) The one exception? John, the veteran, who said, "He's got more problems than a lot of the rest of us." The nurses agreed and, to Harry's horror, rejected his pass request.

"I could not convince them that I could handle it. And that was the most surreal experience. Here I am, I'm in a psychiatric institution and I can't convince them that it's safe to let me go."

He could have walked off the ward anytime and disappeared forever — it was not locked and he was using a fake name — but he felt that he needed to prove to these people that he could handle the real world. He pushed for an easier-to-obtain day pass, which was awarded with ease. Once out, he didn't do anything particularly special, just visited the Stanford campus. He didn't remember if he met with Rosenhan; all he could recall was the feeling of being an alien landing on a parallel version of his home planet. Everything familiar, yet slightly cockeyed.

When he returned to the ward, the staff felt he functioned well enough to be granted an overnight pass and so he took it, spending the night in his own bed (the simple pleasures) next to his wife. Again, he could have left then, never to return, but he felt that he had to see it through to the end. "I would have felt somehow kind of like I was deserting the place in a way," he said.

Everyone seemed to think Harry had adjusted to his off-ward visits well. He was now approaching the hospital's average length of stay — three weeks — and it was time for him to return to life on the outside. This time Harry wasn't the one initiating his release; it was the staff, who approved his release two days after his overnight. During his discharge interview, his diagnosis of schizophrenia was never discussed (as far as Harry could remember); instead, the hospital staff inquired about his living arrangements, his return to school or work, and asked him to draw up a list of people who might be able to help him if an emergency arose again. He reassured them that his environment would be supportive. No drugs were discussed, though they did suggest follow-up therapy. The hospital seemed dedicated not only to discharging Harry, but also to ensuring that he remained well.

Harry was emotional when he said his good-byes. "These were vulnerable folks who in general were caring human beings that were showing their feelings much more than what I was used to academically. And it led me to a closeness that I didn't feel as much on the outside, so I think that was part of it, just the emotion. And again, for somebody who was insecure, not sure I belonged to such an elite place as Stanford, to be in this psychiatric institution and to realize that simply keeping it together was significant. And that was pretty major for me at that time in my life."

The last thing Harry wrote in his journal, according to Rosenhan's notes, was: "I will miss it. I will miss it."

23
"It's All in Your Mind"

Harry and I met face-to-face in a chain hotel in Minneapolis, where I was invited to speak about autoimmune encephalitis to mental health advocates. In person, Harry is more frenetic than the measured cadence of his voice over the phone conveys. He moves his body as he speaks and fidgets in his seat, a ball of energy just waiting for the next marathon (he's an avid runner) to exorcise it.

We spoke about the aftermath of the study and his shifting relationship with Rosenhan. At first Rosenhan was enthusiastic about Harry's hospitalization — or that's how Harry saw it. "He gave me the impression that this was something he really wanted me to be heavily involved with, and work with him, and that kind of stuff." But over time Rosenhan withdrew, growing colder and more detached. They stopped discussing the study. Rosenhan distanced himself from his role as Harry's thesis adviser. And then there was only silence.

"I'm waiting. I'm waiting. He's not around. I'm waiting, and nothing ever happened," Harry said.

Harry put the study behind him to focus on his thesis on smoking cessation, wrote his dissertation, had it reviewed, and finished it by August 1972. All the while Rosenhan maintained an uncomfortable distance. By the time "On Being Sane in Insane Places" came out in 1973, Harry had taken a professorship at Iowa State and hadn't spoken to Rosenhan in over a year. He didn't discover that he had been excluded from the study until he read about it in *Science.* "I felt like I kind of had the rug pulled out from under me," he said.

So Harry decided to write his version — it took all of four hours, the draft coming together in a fever. Not a word was edited. In 1976, Harry revealed his identity as the ninth pseudopatient — the only person involved with the study other than Rosenhan to do so in print. Harry wrote that he had no de-individualization and experienced a deep connection with the staff. His hospital facilities, he revealed, were "excellent," with the nearly 1:1 staff-to-patient ratio creating a "benign atmosphere" and a "genuine and caring" environment.

Though Harry felt vindicated because it helped to "set the record straight," his article didn't make the splash he had hoped, partially because the journal in which it was published wasn't as prestigious as *Science* and partially because in the previous two years Rosenhan's study had been embraced so wholeheartedly that it had become gospel. Rosenhan ignored Harry's article (there are no records of his acknowledging it, even privately, and Harry said Rosenhan never contacted him about it).

I handed Harry the notes that Rosenhan had taken on "Walter Abrams" and braced myself for his response. As Harry read aloud, his brows furrowed: "So . . . let's see . . . 'He was admitted and diagnosed with paranoid schizophrenia.' Wrong. It was chronic, undifferentiated schizophrenia. 'He was discharged twenty-six days later.' Wrong. It was nineteen days."

The mild-mannered man had lost his cool.

"Interesting," Harry said, forefinger on his chin as he read. "Okay. What's fascinating to me is that these are some basic factual inaccuracies that, I mean, don't advance anything. There's no reason for it." Harry was released *with* medical advice, not against. Harry didn't leave "in remission."

Harry was not turned away for "three days" and his ward was not "full," as Rosenhan had written. Yet again, Rosenhan was not only editorializing but filling in gaps with outright fabrications.

I also showed Harry some discrepancies with the numbers. In the files, I had found an early draft of "On Being Sane in Insane Places" sent to marshmallow test creator Walter Mischel for review. In this version, Rosenhan had included nine pseudopatients with no footnote, strongly suggesting that he had written this paper before he decided to remove Harry's data. Not only did the tenor and tone of the paper not change with Harry and without him, but, more strikingly, the numbers didn't, either. That means when Rosenhan took Harry's data out of a sample size of nine, *not one number was affected* — not the average length of stay, not the number of pills dispensed, not the amount of time nurses spent in and out of their cage. I'm no math whiz, but I know that if you remove one out of a relatively small sample size of nine, the aggregated data would *have* to change, at least a little.[1] And the numbers Rosenhan used were so

1. Unless all the numbers are the same, and in this case we know they weren't.

specific: In his paper, he wrote that average daily contact with psychiatrists ranged from 3.9 to 25.1 minutes, for example. This upset Harry — and me.

Just as egregiously, I found notes describing Harry's hospitalization that were repeated almost verbatim in the published paper: "Another pseudopatient attempted a romance with a nurse . . . The same person began to engage in psychotherapy with other patients — all this as a way of becoming a person in an impersonal environment." Both of these details came from Rosenhan's notes on "Walter Abrams," his pseudonym for Harry. How could he include this and also claim that he'd removed Harry as a pseudopatient from the study?

If the editors of *Science* had been aware of these transgressions, I doubt that they would have published Rosenhan's paper. Data, even in a softer, journalistic piece, should at the minimum be sound. I have no doubt in my mind now. Rosenhan's weren't.

Still, Harry believed that the study changed his life for the better. He contemplated pursuing a degree in clinical work, but ultimately decided he could save the world by convincing it to quit smoking. He even changed his appearance.

"I grew a mustache," he said and, as was his habit, moved on to another topic without explanation.

"What is the significance of the mustache?" I asked, guiding him back.

"I think being maybe just a little bit less conventional, because I thought of myself as being pretty conventional." With a bit of facial hair, he had transformed himself into the rebel leader he never thought he could be.

"[The study] affected me, you know, deeply, just the whole experience affected me deeply," he said. He told me about his work with the World Conference on Tobacco or Health's planning committee and about his successful push to convince the group to move their conferences from places like Helsinki and Chicago to cities in developing countries, like Mumbai and Cape Town, where smoking rates are increasing, not decreasing, came from his work as a pseudopatient. "[I'm] quiet, introverted," he told me. After his hospitalization, he realized that "if I really believe in something, I will fight for it."

Harry felt it was pretty obvious what happened (and I agree): Harry's data — the overall positive experience of his hospitalization — didn't match Rosenhan's thesis that institutions are uncaring, ineffective, and even harmful places, and so they were discarded.

"Rosenhan was interested in diagnosis, and that's fine, but you've got to respect and accept the data, even if the data are not supportive of your preconceptions," Harry said. "I do also feel pretty certain, and maybe I'm not being fair, that if I had the experience that the others had, I'm pretty confident that I would have been included . . . Clearly he had his idea, his hypothesis, and he was going to confirm that hypothesis."

Rosenhan included a line at the end of the paper that seemed to subtly acknowledge Harry's experience: "In a more benign environment . . . their behaviors and judgment might have been more benign and effective." But it's a line no one quotes or remembers. Instead, Rosenhan did what so many doctors do to their patients in the face of complexity — he discarded any evidence that didn't support his conclusion. And we're all worse off because of it.

The NPR program *It's All in Your Mind* featured Rosenhan on its opening segment, which aired on December 14, 1972, shortly before the publication of his paper. In the wake of my conversations with Harry, knowing how much gray area Rosenhan had been confronted with, it's infuriating to hear the blind confidence in Rosenhan's voice on tape.

The forty-five-year-old recording opens with a distant trilling of bells. A tribal drumbeat builds into an angry roar. The bells grow louder, louder, louder until a man's voice interrupts: "Psychology, exploring the human psyche. It's all in your mind."

It's a total rip-off of the *Twilight Zone* theme song, which I guess is appropriate, since the radio show I'm about to listen to has a kind of upside-down quality. Hearing the voice of the man whose work you've spent years of your life struggling to understand, yet have barely ever heard speak — a man you once admired but now suspect may have engaged in serious foul play — does feel like being stuck in a room full of books without any reading glasses.

In the twenty-minute interview, Rosenhan walked the host through his experience as a pseudopatient, rehashing his hospitalization and adding a few details that I happened to know were exaggerated, like when he implied that his hospitalization lasted several weeks instead of nine days. "We were administered better than five thousand pills," he said. (In the study, he claimed that two thousand pills were dispensed.)

Interviewer: Do you think that patients can get better going into institutions today as they are in this country?

Rosenhan: No. They were not in any way therapeutic institutions. When you're treating people like lepers, when you can't affiliate with them, when you can't sit down and have a conversation with them, when your bathroom, if you will forgive me, is separate from theirs, and your eating facilities are separate from theirs, and your space is separate from theirs, in no way can you conceive that the half hour that you may spend with them once or twice a week is going to overcome all of that and make their lives better. By and large I think that psychiatric hospitals are non-therapeutic, and would look forward to their being closed.

In disregarding Harry's data, Rosenhan missed an opportunity to create something three-dimensional, something a bit messier, but more honest — instead, he helped perpetuate a dangerous half-truth that lives

409

on today. *I would look forward to their being closed.* Had he been more measured in his treatment of the hospitals, had he included Harry's data, there's a chance a different dialogue, less extreme in its certainty, would have emerged from his study, and maybe, just maybe, we'd be in a better place today.

24
SHADOW MENTAL HEALTH CARE SYSTEM

Decades after the study, Harry returned to a psychiatric hospital — this time as a parent, not as a patient. His daughter Elizabeth was sixteen when she had her first hospitalization for major depression, anorexia, and bulimia (which distracted from an underlying disease that would take another ten years to diagnose — a rare connective tissue disorder called Ehlers-Danlos syndrome). She said that during this hospitalization she felt more like a prisoner than a patient, as if she had done something criminally or morally wrong. "It still gives me that creepy, crawly feeling of being locked in," she told me. There she was heavily medicated and "got so numb that I didn't care anymore." Unlike in the 1970s when her father was a patient, there were no sing-alongs, no ward votes on day passes, no meaningful emotional bonding moments between patients. Just take your meds, watch the TV, and stay quiet until you were "stabilized" enough to leave. Harry couldn't believe what he saw when he visited his daughter. How had his experience decades before been so much more . . . sophisticated? Once she was released back into the care of her doctor, she tapered off the meds. She's still not sure what happened. All she knew was that she needed help but she wasn't sure that the hospital was the right place to provide it.

Meanwhile, Harry's U.S. Public Health Service Hospital followed in the footsteps of Nellie Bly's Blackwell Island — it too was abandoned for decades until it was recently developed into luxury apartments. The Zuckerberg San Francisco General Hospital (as it has since been renamed), where Harry almost ended up, still treats psychiatric patients, but you'd never find people sitting in circles singing "Puff the Magic Dragon" there. There are too few psych beds for too many bodies. Only extreme cases — like a woman who bit off her own finger because the voices told her to — get quick care. "This is the sad part of this work. People so psychotic they can't even get to the hospital without doing something terrible to themselves," nurse manager Jean Horan told the *San Francisco Gate* in 2006. Conditions have gotten so dire that in 2016 dozens of nurses, doctors, and other health care workers protested, saying that the psychiatric unit was in a "state of emergency." Former Bay Area ER psychiatrist Dr. Paul Linde described the revolving-door policy in 2018: "You've got your chow, you've got your shower, you've got your medication, you've got some sleep and now it's time to get out the door."

Patients are often taken by ambulance to emergency rooms, where they are boarded in general hospitals that lack psychiatric care. The hospitals then can't discharge their patients to psychiatric facilities because more often than not, there are no beds available. It creates a logjammed system that fails everyone, as movement is stymied in almost every direction except to the streets or to jails and prisons, also known as "the beds that never say no," said Mark Gale, criminal justice chair of the National Alliance on Mental Illness (NAMI). "These are the choices we are making as a society, because we refuse to fund the completion of our mental health system."

The US is a minimum of ninety-five thousand beds short of need. It's now harder to get a bed in New York City's Bellevue Hospital than it is to land a spot at Harvard University, wrote advocate DJ Jaffe in his devastating 2018 book *Insane Conse-quences*. Sixty-five percent of the non-urban counties in the United States have no psychiatrists and nearly half lack psychologists, too. If the situation continues as it is, by 2025, we can expect a national shortage of over fifteen thousand desperately needed psychiatrists as medical students seek higher-paying specialties and 60 percent of our current psychiatrists gray out.

It's safe to say that Bill Underwood, Harry Lando, David Rosenhan, and presumably the rest of the pseudopatients would never be hospitalized today. If you did have access to decent psych care — not a given in wide swaths of this country — you'd face the following (welcome and necessary) obstacles: "One or more nurses would take vital signs, complete a brief exam and gather some of the patient's history. At least one emergency physician would repeat the process . . . The emergency physicians might order a CT scan of the head or other imaging, depending on the patient's history . . . A psychiatrist would review the patient's chart and any available electronic records . . . From start to finish, these evaluations can take hours," wrote Stanford psychiatrist Nathaniel Morris in the *Washington Post*.

A less welcome reality is this: Most states require that for a person to be hospitalized, she would need to pose a threat or be so gravely disabled and, according to a psychologist, "so disorganized that she would just stand in front of the facility, wander aimlessly in the street, or perhaps stand in the middle of a busy street, with no notion of how to get food or lodging for herself."

One psychiatric nurse laid out what it takes to get care. Ironically, just as it had with Rosenhan and his pseudopatients, it requires acting to get admitted — but follows an entirely different script. In the emergency department, "when being assessed, say (regardless of the truth): 'I am suicidal, I have a plan and I do not feel safe leaving here. My psychiatrist asked me to come here for admission for personal safety, feeling I am a grave danger to myself.' That statement get[s] you back to the psychiatric [emergency department]. Once there, you get interviewed by the psychiatric triage nurse. Repeat the same statement." Only once past these various gatekeepers, onto the psych floor, and in a bed can the patient start telling people what is truly wrong.

In fact, the horror show that is our mental health care system today makes Rosenhan's critiques seem obsolete. "It shows just how quaint the study is — and how misguided it is in a funny way . . . Psychiatry [was seen] as the arm of the state, when in fact [it is] just as much of a victim of the larger relationships of power," said psychiatrist and historian Joel Braslow during an interview.

"It's on the other end of the spectrum today," added Dr. Thomas Insel, former director of the NIMH. "You have people who really do need help who don't get it because there's no place for them to go."

A 2015 study published in *Psychiatric Services* unintentionally imitated Rosenhan's study when a team of researchers posed as patients and called around to psychiatric clinics in Chicago, Houston, and Boston trying to obtain an appointment with a psychiatrist. Of the 360 psychiatrists contacted, they were able to obtain appointments with only 93 — or 25 percent of the sample. (This says nothing of the wait time required for the appointment, nor of the care they would — or would not — receive.)

Dr. Torrey, who founded the Virginia-based Treatment Advocacy Center dedicated to "eliminating barriers to the timely and effective treatment of severe mental illness," said it directly: "People with schizophrenia in the United States were better off in the 1970s than they are now. And this is really something that all of us in the United States are responsible for."

When the promises of community care — first championed by JFK — never materialized, thousands of people were turned out from hospitals (where some had spent most of their lives) and had nowhere to go. When Rosenhan conducted his study, 5 percent of people in jails fit the criteria for serious mental illness — now this number is 20 percent, or even higher. Nearly 40 percent of prisoners have, at some point, been diagnosed with a mental health disorder and their most common diagnoses (some people have more than one disorder) are major depressive disorder (24 percent); bipolar disorder (18 percent); post-traumatic stress disorder (13 percent); and schizophrenia (9 percent). Women, the fastest growing segment of America's inmate population, are more likely to report having a history of mental health issues.

These figures also disproportionately affect people of color who "are more likely to suffer disparities in mental health treatment in general, which results in their being more likely to be ushered into the criminal justice system," said Dr. Tiffany Townsend, senior director of the American Psychological Association's Office of Ethnic Minority Affairs.

There are, at last count in 2014, nearly ten times more seriously mentally ill people who live behind bars than in psychiatric hospitals. The largest concentrations of the seriously mentally ill reside in Los Angeles County, New York's Rikers Island, and Chicago's Cook County — jails that are in many ways now de facto asylums. As someone who knows what it's like to lose her mind, the only worse place than a jail I can imagine is a coffin.

"Many of the persons with serious mental illness that one sees today in our jails and prisons could have just as easily been hospitalized had psychiatric beds been available. This is especially true for those who have committed minor crimes," said University of Southern California psychiatrist Richard Lamb, who has spent the bulk of his half-century career studying and writing about these issues.

This is the current state of mental health care in America — the aftershock of deinstitutionalization, which some call trans-institutionalization, the movement of mentally ill people from psychiatric hospitals to jails or prisons, and others call the criminalization of mental illness. Whatever term you want to use, experts agree that what has resulted is a travesty.

"A crisis unimaginable in the dark days of lobotomy and genetic experimentation" (Ron Powers in *No One Cares About Crazy People*); "one of the greatest social debacles of our times" (Edward Shorter, *A History of Psychiatry*); "a cruel embarrassment, a reform gone terribly wrong" (the *New York Times*).

Though some credit the rise in the mentally ill population behind bars to the fact that America has the highest incarceration rate in the world and to policies like mandatory minimum sentencing and three-strike laws, it's clear that whatever the cause, the fallout has been disastrous. "Behind the bars of prisons and jails in the United States exists a shadow mental health care system," wrote University of Pennsylvania medical ethicist Dominic Sisti. People with serious mental illness are less likely to make bail, and they spend longer amounts of time in jail. At Rikers Island, which is in the process of shuttering, the average stay for a mentally ill prisoner was 215 days — five times the inmate average. Jails are now holding people the same way asylums did in Nellie Bly's era. The ACLU filed a lawsuit against Pennsylvania's Department of Human

Services (DHS) on behalf of hundreds of people who had been declared incompetent by the court. Problem was, there were no beds available, so they were left in jails — in one case in Delaware County, Rosenhan's old stomping grounds, a mentally ill person, too incompetent to stand trial, languished in jail for 1,017 days. The lawsuit's lead plaintiff is "J.H.," a homeless man who spent 340 days in the Philadelphia Detention Center awaiting an open bed at Norristown State Hospital for stealing three Peppermint Pattie candies.[1] During that time, "J.H." had a greater chance of becoming a victim of assault and sexual violence — all because he was too sick to go to trial. In March 2019, the ACLU took the DHS back to court after it "failed to produce constitutionally acceptable results, with some patients remaining in jails for months at a time."

Depersonalization, something about which Rosenhan wrote extensively, is a key feature of prison life. Prisoners are given uniforms, referred to by their numbers, lack even the

1. Coincidentally, David Rosenhan went undercover at Norristown State Hospital as a pseudo-patient in 1973 after "On Being Sane in Insane Places" was published.

most basic privacies, and live without many personal belongings. It's a place where the most valuable currency is to be viewed as powerful, and where the mentally ill are seen as inherently "weak." Prisons and jails are places with "degradation ceremonies" and "mortification rituals." They are not meant to be healing environments; rather, they are punitive, depriving ones.

In Arizona, men, "often nude, are covered in filth. Their cell floors are littered with rancid milk cartons and food containers. Their stopped-up toilets overflow with waste," wrote Eric Balaban, an ACLU lawyer who chronicled his visit to Maricopa County Jail's Special Management Unit in Phoenix in 2018. In California, "Inmate Patient X" at the Institution for Women in Chino in 2017 was not given medication despite being listed as "psychotic," and, after being ignored in her cell after screaming for hours, ripped her own eye out of her skull and swallowed it. In Florida, Darren Rainey was forced into a "special" shower by prison guards. The shower's temperature climbed to 160 degrees, which peeled his skin off "like fruit rollups" and killed him. In Mississippi, "a real 19th century hell hole," non–mentally ill prisoners sell rats to the mentally ill prisoners as pets. In the same place, a man was reported fine and well for three days after he suffered a fatal heart attack. And in the shadow of Silicon Valley, a man named Michael Tyree screamed out "Help! Help! Please stop" as he was beaten to death by prison guards while awaiting a bed in a residential treatment program.

It all reminds me of Erving Goffman's *Asylums,* one of the key texts that inspired Rosenhan's study. Goffman was the sociologist who went undercover at St. Elizabeths Hospital and argued that what he saw there was a "total institution," no different from prisons and jails. He cited examples: the lack of barriers between work, play, and sleep; the remove between staff and "inmate"; the loss of one's name and possessions. Remember Philippe Pinel, the man credited with introducing the concept of moral treatment? In 1817, his mentee Jean-Étienne-Dominique Esquirol described the conditions that led to their enlightenment: "I have seen them, naked clad in rags, having but straw to shield them from the cold humidity of the pavement where they lie. I have seen them coarsely fed, lacking air to breathe, water to quench their thirst, wanting the basic necessities of life. I have seen them at the mercy of veritable jailers, victims of their brutal supervision. I have seen them in narrow, dirty, infested dungeons without air or light, chained in caverns, where one would fear to lock up the wild beasts."

Today, it's worse. We don't even pretend the places we're putting sick people aren't hellholes.

"It's true that the *hospitals* have mostly disappeared," wrote Alisa Roth in her 2018 book *Insane.* "But none of the rest of it has gone away, not the cruelty, the filth, the bad food, or the brutality. Nor, most importantly, has the large population of people with mental illness who are kept largely out of sight, their poor treatment invisible to most ordinary Americans. The only real difference between Kesey's time and our own is that the mistreatment of people with mental illness now happens in jails and prisons."

And then there's therapy — or the farce that passes for it in many prisons. Treatment is often rare and typically revolves around medication management. When therapy does occur in certain jails in places like Arizona and Pennsylvania, it involves doctors or social workers speaking to patients through the metal slats in closed cell doors or, in one egregious case, merely handing out coloring books, wrote Roth.

"Prisoners are under a tremendous amount of stress, and they feel a tremendous amount of pain, and they're not encouraged to think about that. In fact, there's an incentive not to think about it or talk about it, because nobody is interested in it," said Craig Haney, a psychologist who studies the effects of incarceration, whom you may remember as the graduate student who turned down David Rosenhan's invitation to go undercover as a pseudopatient when he was at Stanford.

The culture of distrust goes both ways. During her first day of training in an Arizona state prison, Angela Fischer, a health care provider who later testified as a whistleblower, heard this joke relayed to her by a Department of Corrections employee.

"How do you know when a patient is lying?" the person asked her. Without waiting for an answer, he continued: "Their lips are moving."

Many guards grapple with threats (real or imagined) that the inmates are malingering (or faking) because they want out of a bad situation in the general population or feel they'll get a cushier housing assignment. Though malingering does occur, David Fathi, director of the ACLU's National

Prison Project, said that it is not as common as it's portrayed. More often, people are underdiagnosed and mismanaged: "I mean people who have documented histories of mental illness going back to when they were nine, they get to prison and suddenly they're not mentally ill, they're just a bad person."

Craig Haney agreed, adding that there's no real incentive to lie and game the system: "What's the secondary gain? The secondary gain is that they get taken out of one miserable cell and put into another one that is usually more miserable. If they put you in a suicide watch cell — then you're in an absolutely bare cell with no property whatsoever, sometimes you're in a suicide smock and sometimes they take all of your clothes away and leave you there naked." It reminds me of the second part of Rosenhan's study, when he told a hospital that he sent pseudopatients but never did. Doctors were primed to see pseudopatients everywhere; similarly, prison guards today are trained to see fakers everywhere.

Dr. Torrey, the psychiatrist who warned me that it's worse today than it was during Rosenhan's time, does have some solutions. The Treatment Advocacy Center, which he founded, advocates for adding more beds across the board — in state hospitals and forensic settings — which would reduce wait times and get people out of jails and into proper treatment quicker. Advocate and author DJ Jaffe, Torrey's mentee, a self-described "human trigger warning" and executive director of the Mental Illness Policy Organization, pushes for the implementation of more mental health courts, where judges can divert people with mental illness into appropriate housing and treatment before they've been absorbed into the prison system. He also backs the use of crisis intervention teams made up of law enforcement officers, with the assistance of psychiatric professionals, trained to identify and deal with people with serious mental illness. On the more controversial end, Jaffe has written extensively about the necessity of using legal force to get people to take their meds (something called Assisted Outpatient Treatment), pointing out that many people with serious mental illness don't know that they're sick (a symptom called anosognosia), and for civil commitment reforms so that more people can be hospitalized against their will before tragedy strikes. He and Torrey have both made the case that though the vast majority of people with serious mental illness are no more violent than people without mental illness, studies have shown that a small subset of people, who are typically untreated, are more violent. To those who say these policies infringe on people's civil liberties, Jaffe has responded: "Being psychotic is not an exercise of free will. It is an inability to exercise free will." (I agree that I had zero free will when I was psychotically ill, but I have to admit that it's hard for me to reconcile this perspective with the rest of my experience and misdiagnosis, especially when I think about how many psychiatrists may not be deserving of the power necessary to fully enact these policies.)

Some prisons and jails, resigned to the brutal reality, have implemented changes to reflect their true roles as society's mental health care providers. Sheriff Tom Dart of Chicago's Cook County jail, where a third of the 7,500 prisoners struggle with mental illness, has become a standard-bearer in doing the best with an untenable situation. "Okay, if they're going to make it so that I am going to be the largest mental health provider, we're going to be the best ones," he told *60 Minutes* in 2017. "We're going to treat 'em as a patient while they're here." Cook County provides medication management, group therapy, and one-on-one visits with psychiatrists. Sixty percent of the staff has advanced mental health training, and the jail warden is a psychologist.

But we need money to enact real change. Without the proper allocation of funds, we punish people three times: disinvesting from resources to support them in the first place, arresting them when they exhibit problematic behavior, and then hanging them out to dry when they reenter the community. The system remains broken, and the people who are sickest continue to be ignored and forsaken.

"If I told you that was the case for cancer or heart disease, you'd say no way, we're not going to send people who have freshly diagnosed pancreatic cancer to jail because there's no place to put them while they get treatment," said Dr. Thomas Insel, former head of the NIMH. "But that's exactly the situation we're facing."

25
THE HAMMER

I got a tip to phone Swarthmore psychology professor and social constructionist Kenneth Gergen, who had a close relationship with Rosenhan during his time at Swarthmore. I shared with him what I knew about the study, about Rosenhan's involvement, and about my inability to pin anything to the ground.

He interrupted my ramblings.

"To meet [Rosenhan] and talk with him, he was almost charismatic. Nice, deep voice with a very personable way of relating to people. He was a good networker. He kind of knew people who knew people, and he played the network. He was an excellent lecturer. I mean he just had a certain drama about him. But . . . a number of us in the department, and it wasn't me because I was kind of a friend of his . . . people would say, 'He's a bullshitter.' "

Then he laid the hammer down: "If you've

[only] got one or two examples of things that really happened as they are written in the paper, then let's assume most of the rest was made up."

I hung up the phone and sat still for a moment to take in his words. Could there be any truth to Kenneth Gergen's offhand remark? Exaggerating findings and altering data to fit his conclusions were troubling enough, but inventing people out of whole cloth? That was inconceivable.

Or was it?

People I interviewed kept mentioning one young woman — with beautiful hair, they always added — who worked as Rosenhan's research assistant as an undergraduate at Swarthmore and then continued on in the same capacity at Stanford. If anyone had the answers, they told me, it would be that student. Luckily, Bill remembered her first name: Nancy. With a few educated guesses about the year she graduated, I tracked down a Swarthmore alumni Flickr page, and there, among the middle-aged revelers, I found a picture of a striking woman with long gray hair. She looked straight at the camera, her eyes smiling but her lips sealed, as if flirting with the camera, saying, *You got me.* Her name listed on the page: Nancy Horn.

Over the next few months, I spoke to Nancy Horn four times. We discussed her eclectic work as a therapist, which combines different treatment approaches. We discussed her son, who lives with serious mental illness and has spent time hospitalized and homeless. She regaled me with stories from her undergraduate years at Swarthmore, where she majored in psychology, played volleyball, and met a "charming, witty, incredibly smart" professor named David Rosenhan. She helped him with his altruism research, corralling the children into the testing trailer, rigging the bowling game so that each child would win or lose. She took on a variety of roles in Rosenhan's life: administrator, teacher (sometimes she would help with his classes), researcher, and friend.

"I think he always made people feel special," Nancy said.

The two had lost track of each other in the final years of his life. She learned of his death from a newspaper or academic journal announcement — she couldn't remember which. But she never stopped thinking of him. "I do think about him often . . . I think he was my biggest influence as a role model of a great psychologist, absolutely, a truly great psychologist. And he was great because he was well read, he was wise, he didn't have some sort of narrow-minded sort of focus for himself. He was open to ideas and smart and definitely cared about people, which is what I got into psychology for."

Down to the real reason I called: the pseudopatients.

She recalled working with two graduate students at Stanford — Bill Underwood and Harry Lando — as their point person. She was the one whom Harry contacted from the pay phone and the one he read his medical records to. She also said she visited both men in their hospitals.

"Did Rosenhan prep you for what to look for? Did he say look for certain . . ."

"No."

"He just trusted that you would . . ." As I trailed off, I thought about myself as a recent college grad. I would never have been responsible enough to monitor someone's mental health. (I'm still not.) Wise beyond her years, Nancy devised a method to examine them for any signs of distress. She looked at their speech patterns, asked them what they were doing to pass the time, inquired about their medications, made sure they weren't emotionally unstable.

"I looked for fifty things at once," she said. "You know, if you're in a crazy situation, it can be crazy-making. So you have to be sure somebody isn't going nuts from being in the situation."

Despite Nancy's maturity, the fact was that Rosenhan left a huge responsibility in an assistant's hands. This was at best unprofessional. Forget the paper's transgressions: Even if the study's data were flawless, there's no way the Institutional Review Board (IRB), which oversees academic research to protect "the rights and welfare of human research subjects," would approve it today. It would pose too great a danger to participants, hospital patients, and, to some degree, its research assistant.

But what of the other six pseudopatients — the Beasleys, Martha Coates, Carl Wendt, and the Martins?

Nothing. Despite their closeness, Rosenhan had kept the identities of all but two of the paper's pseudopatients hidden from even his research assistant. I walked her through the information that I had gathered from Rosenhan's notes and manuscript. I told her about Sara Beasley, #3, who almost swallowed her medication to drown out her anxiety, and Robert Martin, #6, the pediatrician who developed paranoia about his food.

"He thought his food was being poisoned? That's not good."

"No. And if you heard that, you would have said we need to get him out, right?"

"Oh, are you kidding? Oh my God, he'd be out in a heartbeat. That would be ridiculous. Oh, I'd be so upset if I heard that," she said.

When I described Laura Martin, #5, the artist, she asked, "That was the one at Chestnut Lodge?"

"Chestnut Lodge?" Listening back to my recording, I hear my voice pitch up. This was a solid lead at last, a counterweight to Kenneth Gergen's suggestion that they might be made up. Chestnut Lodge was a famous private psychiatric hospital located in the shadow of DC, where Washington's "eccentric" elite lived out their unraveling in style. Two popular novels, *I Never Promised You a Rose Garden,* which author Joanne Greenberg based on her hospitalization there and treatment by famous in-house psychotherapist Frieda Fromm-Reichmann, and *Lilith,* the story of a relationship between a patient and her attendant written by a former employee of the hospital, were huge bestsellers that were made into movies. Chestnut Lodge, I would find, was a dividing line in the battle between the brain and the mind — the place from which psychoanalysis sent its final flare.

Chestnut Lodge was founded on the idea that the asylum should be a safe place for a rich person to live with dignity. The end goal wasn't really a "cure" per se; instead, patients would spend years (in some cases, a lifetime) on the gorgeously manicured grounds, going from tennis to art therapy and, of course, to daily talk therapy. The hospital did not partake in the hideous treatments that checkered psychiatry's other venerable institutions — no lobotomies, insulin coma therapies, or electroshock here — and it even put off prescribing drugs. And then came Dr. Ray Osheroff, a depressed forty-one-year-old kidney specialist, who was admitted to Chestnut Lodge in 1979 and was diagnosed with "narcissism rooted in his relationship with his mother." The Lodge employed "attack therapy" and "regression therapy" over the course of nearly a year, which only worsened his condition, as he lost forty-five pounds and paced almost constantly, upward of eighteen hours a day. Osheroff's parents intervened and moved him to a more traditional psychiatric hospital, where he was diagnosed with depression, treated with antidepressants, and released nine weeks later. Osheroff sued Chestnut Lodge for malpractice (settling out of court for a rumored six figures) — a case that became about something larger than Osheroff, by proving that "psychiatry was a house divided," said Dr. Sharon Packer in a belated obituary for Osheroff written in 2013. "The hallowed walls of psychoanalysis were tumbling down."

Knowing all of this history, it is hard to believe how little Chestnut Lodge left behind. The Lodge's heyday ended without fanfare when the veritable institution filed for bankruptcy in 2001 and the property was sold to luxury condo developers. Then, on July 13, 2009, the barking of an "aggravated dog" alerted the neighborhood that the historic building had gone up in flames. Everything was lost. Chestnut Lodge had hardly left a footprint.

But there are some among the hospital's former employees who keep the memory of Chestnut Lodge alive. One psychologist, who also works at the NIMH, brought a scrapbook of pictures from her time there during our first interview. (How many former employees keep pictures from their old jobs — especially jobs located at psychi-atric hospitals?)

"This is a summertime photo. See? The grounds are beautiful," she said, pointing to the chestnut trees. She showed me the gym, and the pool, and recalled the time when a wedding party wandered among the trees in search of the perfect photo op, without re-alizing they had stepped onto the grounds of a psychiatric hospital. The psychologist had felt so proud, even as she shooed them away, to know that the setting was as beauti-ful and peaceful to outsiders as it was to her. "Please be kind to Chestnut Lodge," she said to me. "I really loved it."

I told her about my mission — about the Rosenhan experiment, about tracking down the pseudopatients, about the possibility that one of the undercover agents had infiltrated the Lodge. She had not heard about the study happening at Chestnut Lodge, but she admitted that it was long before her time. Luckily for me, when the hospital was being dismantled and picked apart after its bankruptcy and before the fire, she had squirreled away a metal filing cabinet that held the hospital's patient records — three-by-five cards printed with the names of and information on each patient who had visited the Lodge since its inception: length of stay, diagnosis, dates of admission and release — files that would have been thrown out without her intervention. I was thrilled: This would be more than enough to find my pseudopatient. She agreed to see if anyone matched the artist's description and length of stay, though she refused my offer to help her dig, citing patient privacy laws.

She went her way, and I went mine.

If Chestnut Lodge checked out, and I could find Laura Martin, pseudopatient #5, I would feel somewhat better about the whole enterprise. There was hope: Rosenhan had visited DC for six days in 1971 — smack-dab in the middle of the study, and likely the time when Laura Martin, the only pseudopatient who attended a private hospital, went undercover — so it was possible that he had visited Chestnut Lodge then.

I returned to Rosenhan's manuscript to study the parts about Laura, the famous abstract artist who was hospitalized for fifty-two days and the only subject to receive the diagnosis of manic depression. I reread a chapter in his unpublished book that recounted the time Rosenhan was summoned to Laura's hospital to consult on an "interesting case," only to discover it was his own pseudopatient.

Rosenhan took detailed notes on the case conference of his pseudopatient, quoting Laura's psychiatrist, who used florid terms to diagnose her using her paintings — "The ego is weak," the doctor said as he examined one of the six works she created at the hospital. Despite his gross misjudgment, Laura did use the opportunity to do some work on herself, and, in Rosenhan's description of that process, the outline of a woman and an artist emerges. Rosenhan described the worries that she faced about her pediatrician husband, whom she feared was working himself into an early grave (this was Bob, who in his own time as a pseudopatient would obsess about the food); her concerns about the younger of her two sons, Jeffrey, who had begun experimenting with marijuana; and her issues with honing and maintaining her creativity as a painter.

I then asked dozens of people who knew Rosenhan if he had befriended any famous female artists in his lifetime, but no one had any solid suggestions. I made lists of famous female abstract artists from that era and phoned art historians. They floated several names: Anne Truitt, Joan Mitchell, Mary Abbott, Helen Frankenthaler, all dead ends. The National Museum of Women in the

Arts in Washington sent me a list of books. There were false positives. The mother of one of Rosenhan's Swarthmore students was a pretty famous sculptor. No dice. I emailed Judith W. Godwin, an abstract artist from New York whose work hangs in the Metropolitan Museum of Art. She responded to my email with the kind but firm: "I didn't take part in this study. Good luck in your research."

And then a hit.

Grace Hartigan, who was born in Newark in 1922 and died in 2008, began her career as a draftsman in an airplane factory. With no formal training, she started re-creating Old Masters. In the 1960s, she incorporated images from popular culture into her intensely colorful work — an early version of pop art. "I didn't choose painting," she said. "It chose me. I didn't have any talent. I just had genius."

She married four times. Her first husband's name was Bob. Ding, ding, ding! Only issue: Her first husband was out of the picture by 1940. Not so ding, ding, ding-y. Here's where I wanted to take a victory lap: Her fourth husband, Dr. Winston Price, an art collector whom she wed in 1960, was a famous epidemiologist from Johns Hopkins University, obsessed with finding the cure for the common cold. There was little he wouldn't do for his research — he even injected himself with an experimental vaccine for viral encephalitis, which gave him spinal meningitis, starting a decline that went on for a decade until his death in 1981.

Could these two have been my Martins? Winston Price put his life on the line for his work, so admitting himself to a psychiatric hospital wouldn't be a stretch. Grace, who struggled with her own demons, including alcoholism, had a vested interest in the study of madness and its overlap with creativity. It seemed plausible to me and to Grace's biographer, Cathy Curtis, who added this: Grace Hartigan's son Jeffrey (same name as Laura's son, according to Rosenhan's notes) had life-long issues with drug abuse, just as Laura worried that her son Jeffrey smoked marijuana. But Cathy tempered my confidence. "At the tender age of 12 [Grace Hartigan] sent him to live with his father in California. She really had very little to do with him the rest of his life," Cathy said. "She would tell people she hated him."

"On the bright side," Cathy added in an email, "if I had to quantify the Hartigan possibility I'd say 80 percent."

Eighty percent. I'd take those odds. Yes,

Grace had only one child, not two, and likely didn't care enough for the child to worry about him, but these are things that could have been exaggerated or miscommunicated to or by Rosenhan. To shore up the possibility, Cathy recommended I contact Grace's longtime assistant Rex Stevens, who worked with her for twenty-five years.

"It's not Grace." This was Rex Stevens. He said it with such authority that it felt like a shove. The timeline was wrong, he said. The description of her painting was wrong. The relationship with her art, with her husband, with her son — all wrong. But the most damning part from his perspective? She would have told Rex.

"I know everything about her," he said.

I brushed this phone call off as a product of resentment. I'm sure I would have been dismissive if I heard that someone I knew for that long was hiding something this big. I contacted a researcher at Grace's archives at Syracuse University, which contained twenty-five linear feet of correspondence, notebooks, and diaries from the bulk of her career. But the researcher could not find one letter to or from David Rosenhan. The odds were dropping precipitously that Grace Hartigan was my Laura Martin.

Pseudopatients #5 and #6 were still

unaccounted for.

During one of my earlier research trips in Jack's condo, when I had stumbled across the outline of Rosenhan's unpublished book with handwritten notes that eventually led me to Bill Underwood, two other unexplored clues had intrigued me, too. Beautiful but almost indecipherable letters (that took the help of both Florence and Jack to decode) spelled out: "Letter from Leibovitch"; above it: "Psychotherapy — use letter from Cincinnati."

I had kept them in the back of my mind as possible leads, but I didn't know how to connect them until I stumbled on a series of letters written by a woman named Mary Peterson nestled in a draft of the sixth chapter of Rosenhan's unpublished book. The letters detailed Mary Peterson's experience at Jewish Memorial Hospital in Cincinnati.

Cincinnati.

One letter to Rosenhan from Mary Peterson described the twelve days that she spent on Jewish Memorial's psychiatric ward. Mary narrated the story of her hospital stay into a recorder and sent the tapes to Rosenhan, who asked his secretary to transcribe them. The transcriptions, which were only partially completed, detailed an extensive cast of characters, among whom quite a few names matched the descriptions of patients in Rosenhan's notes on Sara Beasley, pseudopatient #3. Mary and Sara also had similar descriptions of their first anxious night on the ward, when Sara almost swallowed her pills.

"Got one of them!" I scrawled in the margins of my notes.

Rosenhan had kept Mary's envelope, so I had her address, which helped me track her to Cleveland, Ohio — only to find out that she had recently passed away and that her husband (named John, also the name of the husband of Sara, pseudopatient #2) predeceased her. Her obits made it clear: This woman was filled with vigor. I read through the cooking columns she had written for the local paper and ordered her self-published book of adoring short stories about living in the Queen City. "An angel on wheels" is how a local writer described Mary Peterson, who was often spotted on her pink bicycle. "Sometimes I think there are angel wings protruding from her back when I see her biking!" Now I was doubly disappointed — not only would I never get a chance to ask her about the study, but I would never get to meet this remarkable woman.

But my excitement masked some problems. First, Mary Peterson would have been too young to fit Rosenhan's description of someone "gray-haired" and "grandmotherly" in 1969. Mary's occupation — an economics professor instead of an educational psychologist — didn't fit. Mary Peterson's stay in the hospital was longer than Sara Beasley's, another issue. Her husband, though also named John, was an architect, not a psychiatrist. But perhaps Rosenhan had changed autobiographical details to preserve identities, as we'd now seen him do to some degree with both his own and Bill's records (though age, occupation, and physical description remained intact). Why else would these letters be filed with drafts of his unpublished book?

A quieter issue was that Mary Peterson spoke to Rosenhan about her long-standing history of depression and anxiety. She confessed as much in her letters, telling Rosenhan that she had spent the last decade on tranquilizers and regularly saw a shrink. Would Rosenhan have sent a woman with a history of mental health issues in as a pseudopatient?

But the hardest fact of all to assimilate

445

into the narrative was the timing: If her notes were correct, Mary was admitted to Jewish Memorial Hospital in 1972, right around the time Rosenhan handed in his first draft of "On Being Sane in Insane Places" to *Science,* making it impossible that she was Sara Beasley, the third pseudopatient who helped kick-start the study in 1969.

I contacted Mary's surviving sister and childhood best friend. Neither recalled any mention of the study. Neither had heard of David Rosenhan.

Finally, I shared the letters with Florence, the keeper of the files, to get her take. With her clinical eye, honed from years as a psychologist at an acute care facility and working with the "worried well" in her private practice, I believed her when she concluded: "There's no way that Mary was a pseudopatient. She was a real patient."

Why, then, did Rosenhan file this inside his unpublished book on the pseudopatients? If the letter had arrived only after his study, I wondered if "use letter from Cincinnati" could mean that he'd planned to supplement his discussions in the book with Mary's experiences. It was a possibility, at least, though he had not yet used any hospitalizations other than the pseudopatients' in the existing drafts of the book.

In addition to Mary's letters, Rosenhan kept among his notes two journals: the first, one-hundred-plus pages from a Swarthmore undergraduate who in the summer of 1969 spent a month at Massachusetts General Hospital observing their psychiatric unit; and the second, unfinished diaries from two Penn State undergraduates, who, following the publication of Rosenhan's study, went undercover in Pennsylvania psychiatric hospitals. Why had Rosenhan kept these files, yet retained none of his own pseudopatients' notes?

More questions — zero answers.

Despite a glimmer of hope, the Beasleys, pseudopatients #2 and #3, and Martha Coates, #4, remained at large.

I naively thought that Carl, the recently minted psychologist whom Rosenhan feared was becoming addicted to the pseudopatient charade, would be simple to track down, thanks to reporting done before me. Several people had suggested that Martin Seligman, considered "the founding father of positive psychology," who coined the term *learned helplessness,* was a pseudopatient. His biography matched up with only one: Carl, my #7. When I reached him and eventually interviewed him, however, he delivered the bad news — he was not Rosenhan's pseudopatient, though he did go undercover at Norristown State Hospital with Rosenhan for two days in 1973 *after* the publication of "On Being Sane in Insane Places" to help Rosenhan gather more color for his book. Medical records I tracked down confirmed this.

So it was back to square one. If Rosenhan's notes were to be believed, Carl's age distinguished him. He seemed to be somewhere between thirty-eight and forty-eight, high for a newbie psychologist who had only recently received his clinical PhD. I knew he wasn't at Stanford because the university didn't offer an advanced degree in clinical psychology, meaning that Carl likely came from another institution, and, let's be honest, that institution could be anywhere on the East or West Coasts (really anywhere in between, too). Even though I now considered Rosenhan to be at best an unreliable narrator, he was the only guide I had. But after hundreds of emails exchanged with anyone ever connected with Rosenhan, hours spent on the phone, and days sorting through his papers and correspondences to find any legitimate clues, I was giving up hope. No one fit the bill — until one finally did.

I kept coming across the name Perry London. *It's too bad Perry isn't here,* people kept saying to me. *He'd know everything.* Rosenhan and Perry worked and played together, co-authoring over a dozen papers, mainly on hypnosis, and writing two abnormal psychology textbooks together. Both were larger than life (though Perry, unlike Rosenhan, was large in stature, too); both had big booming laughs with big booming personalities. Perry would know all there was to know about the study — if anyone did — but he had died in 1992. The past was largely buried and gone until I arrived in the Londons' lives, reopening old wounds in an effort to resurrect a man I'd never met.

His daughter Miv, a psychotherapist in Vermont, responded to my email and connected me with her mother, Vivian London, Perry's ex-wife. I was properly vetted enough for Vivian to Skype with me from her home in Israel. She reminded me of my mother, and not just because they were both born in the Grand Concourse area of the Bronx, but also because they both have tough, take-no-shit exteriors. She shared the origin story of Perry and Rosenhan's long-standing friendship. Vivian had connected them when Rosenhan worked as a counselor at the summer camp that her family owned.

"Everyone loved David," she told me. He was the kind of counselor who could calm any homesick child, curling up beside a particularly upset one and soothing him to sleep. One summer when Rosenhan couldn't attend, he sent a friend of his to take his place as a counselor. The following year this friend couldn't make it and sent another friend in his stead, a boisterous young man named Perry London. Vivian and Perry started a summer fling that led to a wedding that also led to Vivian's introducing Rosenhan to Perry.

When I mentioned "Carl Wendt," my seventh pseudopatient, and the brief description that Rosenhan had included in his notes, Vivian stopped me. "Was he an accountant in Los Angeles?" she asked.

"He may have been."

"That outline kind of matches a good friend of Perry's in Los Angeles."

"What was his name?"

She hesitated. I pressed. She pushed back. For the next five minutes, we debated. *What if he doesn't want to be found?* she asked. *If he kept this secret for this many years, maybe he didn't want to expose it?* I countered, explaining that there was nothing to be ashamed of and if his family wanted him to remain anonymous, I would follow their wishes. Eventually, she relented.

"Maury Leibovitz," she said.

The name sounded so familiar. Vivian told me a bit more about this Maury character: Maury, like Carl, left behind a lucrative accounting gig in his early middle age to return to school and get a doctorate in psychology. He landed at USC, where Perry London became his teacher, mentor, and close friend. It was not implausible (at all!) that Rosenhan would have reached out to Perry for help in finding pseudopatients or that Rosenhan would have met Perry's students during, say, a Friday-night Shabbat party (there were lots of those happening then). There was only one degree of separation between Rosenhan and Maury. And Maury fit the Carl bill to a T. Maury was even a fan of tennis, according to Vivian, which matched Rosenhan's comment in a draft of his book that called Maury "athletic."

When we logged off Skype, Vivian sent me a follow-up email. She was nearly as excited as I was. "It has become obvious to me that Maury is your man. I don't even understand how I could have doubted it."

I put on a pot of coffee, opened up my filing cabinet, which was filled with photocopies of Florence's files, and resumed my dig. I was sure I had seen that name, Maury Leibovitz, before at some point, but I couldn't place it. It didn't take me long to find a reference. In that same outline of his book, marked in pencil — right by the CINCINNATI note that (mis)led me to Mary — Rosenhan wrote the word: LEIBOVITCH.

Did he mean Leibo*vitz*?

It made so much sense. Not only did the two have a friend in common, but Rosenhan, I found, actually wrote a letter of recommendation for Leibovitz in November 1970, which meant that they also had a working relationship. This couldn't be a coincidence, could it?

Maurice ("Maury") Leibovitz wasn't exactly difficult to track down. A Google search yielded a glowing *New York Times* obituary, published the same year Perry died. He was a major figure in the art scene in New York as the vice chairman and president of the Knoedler Gallery (now defunct after lawsuits for fraud long after Leibovitz's death), a New York institution. New Yorkers regularly walk by the Gertrude Stein statue sculpted by Jo Davidson in

Bryant Park, which Maury donated to the city.

With Maury Leibovitz came a theory about how a famous painter — pseudopatient #5, Chestnut Lodge's "Laura Martin" — got involved with the study. Maury Leibovitz was a man deeply embedded in the art world. He could easily have been the bridge between Rosenhan and Laura.

Leibovitz was survived by three sons, an ex-wife, and a girlfriend. Of the sons, Dr. Josh Leibovitz, a Portland-based addiction specialist who had inherited his father's interest in the mind, was the easiest mark. I left a message at his office and waited.

The next day a man's Southern California drawl greeted me on the phone.

"I have reason to believe that your father was one of [Rosenhan's] pseudopatients, one of the volunteers. Does this make sense to you at all?" I asked Dr. Leibovitz.

"Really?" he asked.

"Yes." I could feel my heart jumping up to my throat. Seconds passed before he spoke again.

"No," he said steadily. "I don't believe that is true."

I sighed. Over the course of the next twenty minutes I tried to make my case, which Dr. Leibovitz batted down: Maury

453

would have been too old to be my Carl (Maury was fifty-two, when Rosenhan listed him as anywhere from thirty-eight to forty-eight, depending on the document, though, really, how much could we trust Rosenhan's descriptions at this point?). He also was famously claustrophobic and would never have allowed himself to be confined to a mental hospital. And finally, the family was out of the country in Zurich during the time that the study took place.

"I'm sorry to disappoint you," he said. "But it's not my dad."

But it *was*. It *had to be.* I pushed, positing the delicate question: Could it be possible he didn't know his father as well as he thought he did?

"I've got to tell you, my dad was not a man to keep secrets. We were extremely close, so I doubt he would withhold something like that. I mean, I knew every detail of his life," he said. "My dad would have probably written a book about it. He would not have been quiet about it."

But why, I added, would the name Leibovitz, though spelled wrong, be in Rosenhan's notes? I was like a bloodhound on a scent, and nothing he could say or do would knock me off it. I asked him to speak with his mother — she would have noticed that her husband was absent for at least sixty days (this was another issue with Carl: Some of Rosenhan's documents said he was in for sixty days over three hospitalizations, while others said seventy-six days over four hospitalizations), so to my mind she would be the deciding vote. He promised to get back to me with an answer but denied my request to speak to her directly, effectively asking me not to waste his elderly mother's remaining moments on earth.

At this point, I was clinging to the hope that this would all work out like a Doomsday cult member clings to her belief that the end is nigh even as the sun rises the next morning.

Another setback came that same week, this time in the form of a text message from the Chestnut Lodge psychologist, who had finished going through the hospital's patient files.

"No one with the name or initials [of Laura Martin] was admitted in the late 60's or early 70s." Worse still, no person from 1968 through 1973 stayed at the hospital for *only* fifty-two days. The average stay, even into the 1980s, was fifteen months. "There was no way that this patient and her art work would have been presented during a [fifty-two-day] stay," she wrote. To get a patient conference, you had to be in Chestnut Lodge for *much* longer. Doctors didn't feel they knew their patients well enough to present a whole case study five weeks in. But Nancy Horn had recalled that *someone* had been there. Did she get it wrong or had Rosenhan lied about that, too?

As I was reeling from the news, I received this email from Dr. Leibovitz: "I spoke with mother and she is really confident that my father was never involved in such a study. She is 86 and a very private person. She was not interested in discussing any further. Good luck with the research. Keep me posted if you ever find out who that person was."

Why did every single one of my leads go nowhere? Why had Rosenhan so obscured the path to these pseudopatients? What was he protecting? I felt betrayed by a man I'd never met. Had I spent my time pursuing phantoms in a fictional universe?

I returned to Laura Martin's file one more time, this time with a furious, skeptical eye. I reexamined Rosenhan's description in his unpublished book of the patient conference, where Laura's psychiatrist used her paintings to reveal the underlying symptoms of her mental illness. Rosenhan quoted him directly: "The upper portion of the painting is the patient's wish. Unable to handle the impulse life that surges beneath, she wishes for blandness. And perhaps in her better moments she can mobilize that blandness. But in the main it is difficult. She lacks the ego controls, on the one hand, and the impulses are too strong on the other. The blandness that she desires, representing both peacefulness and absolute control over her impulses, simply cannot be achieved. At best she can achieve moments of calm, punctuated alternately by depression and loss of control."

The psychobabble continued. Her doctor moved on to four other paintings and then arrived at her sixth, and final, one. "The bottom half of the painting [is] much less intense . . . the colors here are better integrated . . . Mrs. Martin's impulse life is better integrated." A thick line separating top and bottom became proof to the doctor that Laura had improved under the watchful eye of his care.

Knowing now how far Rosenhan was willing to stretch truth, the problem here seemed unmistakable. This scene was too on the nose. Even the psychoanalytic interpretation of her paintings sounded clichéd, too much of a *New Yorker* cartoon depiction of a pipe-smoking analyst. And then

457

the unlikely coincidence that Rosenhan himself was consulted on her case — he wasn't a clinical psychologist and hadn't worked with patients since his early days after getting his PhD, so why would someone in Washington, DC, call him to travel *to see one of his own patients*? Then there was the issue of how he managed to pay for these hospitalizations. He wrote in his private letters that he funded the hospitalizations himself (to avoid insurance fraud and other possible illegalities). Fifty-two days in one of the swankiest hospitals in the country would have cost a fortune, even then. Where did he get the money?

Kenneth Gergen may have been right after all. Did *any* of this even happen?

26
AN EPIDEMIC

Now the question was: Had Rosenhan outright invented pseudopatients to up his "n" — or the number of subjects in his data set — to lend more legitimacy to his findings? Had getting away with his exaggerated symptoms emboldened him to go ten steps further and invent pseudopatients? Did he get caught up with a book deal and out of desperation decide to fill in the blank pages? This elaborate ruse no longer seemed impossible: There were Mary Peterson's letters and the undergraduates' journals and their odd placement within Rosenhan's files; there was Chestnut Lodge and his "famous artist" pseudopatient Laura Martin, whose case conference sounded a bit *too perfect;* and then there was Carl, who so closely resembled one of Rosenhan's friends, but one who had never participated in the study.

I hadn't wanted to believe that the man I

had so admired could turn out to be *this* —whatever *this* was. My goal was no longer to just find the pseudopatients; I was now seeking out proof that they didn't exist. So I spent the next months of my life chasing ghosts. I wrote a commentary for the *Lancet Psychiatry* asking for help. I made a speech at the American Psychiatric Association, calling for anyone who had ever met David Rosenhan to contact me. I hunted down rumors, spending a month pursuing a lead that St. Elizabeths Hospital in Washington, DC, was one of Rosenhan's locations, just because the Wikipedia page on his study included an image of the hospital as its key art. I even hired a private detective, who got no further than I had. I contacted everyone who had ever entered Rosenhan's orbit, and was shocked to find as I departed further from his inner circle how many people wanted no part in the retelling of his story, including one former secretary who may have had access to some of his work during the writing of "On Being Sane in Insane Places." When I reached her, all she would offer was, "Well, he did often use some 'creative thinking.' " She laughed, and then her tone darkened: "I have nothing nice to say, so I won't say anything at all."

All navigable roads led back to Bill and

Harry. Students, fellow professors, and friends either knew nothing about the study or led me straight back to the two I had already found.

I researched lying and found a splashy *Daily Mail* article that claimed to offer "scientifically proven" ways to spot a liar using textual analysis that scoured writing for "minimal self-references and convoluted phrases" and "simple explanations and negative language." Unfortunately, when I ran this by a real expert, Jamie Pennebaker, a University of Texas social psychologist who studies lying, he said that it was impossible to suss out a liar from text alone, and that anyone who told me otherwise was probably lying to me.

I ran all of this researched skepticism past Florence. She had often called Rosenhan a "storyteller" and said that he might have been happier as a novelist than a researcher, but would the fantastical side of him go this far? At first, Florence doubted he would. But upon further reflection she wrote me an email:

"I continue to wonder whether some of these folks were fabricated . . . it would certainly explain why David never completed the book."

It was a good point. His publisher,

Doubleday, sued him in 1980 in the Supreme Court of New York to recoup the first installment advance for *Locked Up* (by then he had changed the name from *Odyssey into Lunacy*), which was already seven years late and would never be delivered. Had the editor's encouraging comments, in which he also suggested adding more detail about the "vague" pseudopatients, spooked Rosenhan? To almost everyone I spoke with, his abandonment of the study that made his career was the most concerning, even damning, evidence that something was seriously amiss.

After the publication of "On Being Sane in Insane Places," Rosenhan returned to researching altruism, publishing a paper on the effects of success and failure on childhood generosity. After 1973, he jumped from topic to topic, from mood and self-gratification to the joys of helping to moral character to pseudoempiricism to the study of nightmares experienced after an earthquake. The research all seemed a bit unfocused. In fact, one colleague told me, after all of his success with his famous paper and professorship at Stanford, "David became sort of less involved academically . . . less research oriented generally."

His most successful work, after the study,

was a textbook on abnormal psychology that he published with Martin Seligman and that as of this writing is in its fourth edition and is still used in classrooms around the country. He researched juror behavior, including one paper on how note-taking aids jurors' recall of facts and another on their ability (or, rather, inability) to disregard facts that judges had ruled inadmissible. He also joined forces with Lee Ross and Florence Keller as trial consultants — or psychologists who help with trial preparations, like jury selection and opening and closing statements — early adopters in the use of the social sciences to aid in legal analysis.

His research on "intense religiosity," which friends cite as his most beloved work, though it was never published, found that a shocking percentage of Stanford students believed not only in God (75 percent) but also in creationism (59 percent), leading Rosenhan to conclude that although "for most of this century religiosity was negatively correlated with intelligence and social class, there is increasing evidence that the direction of that correlation has reversed sharply."

Okay, but: As interesting as this sounds, does it sound realistic to you that 59 percent of the Stanford student body believed in creationism into the 1990s?

Perhaps I'm being unfair, my antennae now hyperattuned to any signals of fraud. All of this is to say that after he published his classic work, the study that would help bring psychiatric care as he knew it crashing down, except for a brief follow-up he never again published research on the subject of serious mental illness and psychiatric hospitalization.

His dual professorship in law and psychology, which came along with not only a higher salary than his psychology peers' but also the benefit of two separate offices, afforded him a cloak of invisibility that some of his students and colleagues thought was shady. "Whenever you'd try to find him in the Psychology Department, he would be in his law office," one former graduate student told me. "Whenever you went to find him in law, he'd be back in psychology." He seemed to be everywhere and nowhere.

Eleanor Maccoby, one of the most respected psychologists in her field of developmental psychology, who worked with Rosenhan for forty years and even headed the tenure committee when Rosenhan received the honor, didn't soft-pedal during our interview at her retirement home on the eve of her one hundredth birthday. "I was suspi-

464

cious of him," she said. "Many of us were." When his tenure came up for review, the department was divided, she recalled. Of the study, she said, "Some people were doubtful about it. It was impossible to know what he had really done, or if he had done it." Though they ultimately decided to grant him tenure because of his talents as a lecturer, this doubt shadowed him throughout his professional career. "His reputation gradually shrank away," she said.

Marshmallow test creator Walter Mischel, who passed away in 2018, told me that he didn't have much contact with Rosenhan, despite having edited an early draft of his study. In a private correspondence, however, he was more forthcoming: "I never really connected with Rosenhan, found him a pain when I was chair, and thought he avoided work like the plague. I also was not drawn to his research, and made a point of staying away from it and from him."

I contacted a woman who had loved Rosenhan years ago and still held on to his memory, even if that love had long ago soured. She agreed to speak with me with one caveat: I never ask her about their affair. It was a tough agreement to uphold, especially when she took out a box of recordings of his lectures that she had kept for decades.

"He could make you feel like the most important person in the world, just in the way he talked to you," she told me. She had worked with many psychologists and said that they all shared one common trait. "Look at their focus of study and you can count on it that that's what they have a problem with. That's why they study that particular area."

"Oh, that's funny," I said. "What would Rosenhan's problem be?"

"Well, morality, altruism, being a decent person, I suppose," she said. She was laughing but in a raw way. "I mean, I always used to say, 'He's polishing his halo again.' He had an uncanny ability with all his training on personality and character and so on; he had an uncanny way of projecting himself. That you saw him exactly the way he wanted to be seen."

Rosenhan's research assistant Nancy Horn was one of the holdouts who refused to believe that Rosenhan was capable of such dishonesty. She gave me a resounding "absolutely not possible" when I broached the possibility that he had made up a good deal of the paper. His Swarthmore student Hank O'Karma, the author of one of the undergraduate journals that Rosenhan kept

466

in his pseudopatient files, was adamant that he couldn't have, too. Rosenhan's son, Jack, to whom Florence and I posed the possibility over lunch at a diner in Palo Alto, also dismissed it, and added, "My dad was a storyteller. That's true. But I do not think that he would ever do anything that would mess with this research."

When I presented Bill with the facts, he seemed uncertain. "I don't know," he said. "Seems unlikely to me. It's hard for me to imagine."

Harry disagreed. "I never thought of him as a BS artist as an undergraduate. I felt neglected as a graduate student, but that's a different issue. But this . . . ," he said, referring to what Rosenhan had written about Harry's experience, "is total fiction."

All of the little things — the wig, the lying about his hospitalization dates, the exaggeration in his medical records, the playing with numbers, the dismissal of Harry's information, the unfinished book, his never tackling the subject again — all of these piled up. Rosenhan does not seem to be the man I'd believed in.

It wasn't the first time a paper published in a journal as esteemed as *Science* had been called into serious question, even exposed as an outright fraud. One of the most ignoble examples is social psychologist Diederik Stapel, once famous for his article published in *Science* about a correlation between filthier train platforms and racist views at a Utrecht station. The media hailed the piece. He followed it up by claiming to find a link between carnivorous appetites and selfishness. Then the ground fell out beneath him. The *New York Times* called him "perhaps the biggest con man in academic science." For years he had invented data in more than fifty papers. Diederik Stapel's story, though extreme, revealed not only that this level of con *could* happen but that the environment — where journals select articles that will make splashy news, where there's pressure to leave contradictory data out (something called "p-hacking"), where negative, non-sensational studies go unrewarded and unpublished, where grant money and livelihoods are dependent on publishing (the "publish or perish" issue) — provided a hothouse environment for people like Stapel looking to exploit the system.

Right now the field of psychology — especially social psychology — is in the midst of a "replication crisis," and a few critics have turned their sights onto some of the field's most cited works, from "power posing," to "the facial feedback hypothesis," to "ego depletion." Bryan Nosek from the University of Virginia started the "Reproducibility Project," which repeated one hundred published psychological experiments and could successfully reproduce the findings from fewer than half of them.

Walter Mischel's marshmallow study (the one Bill's daughter took part in at Stanford), in which preschoolers who were able to restrain themselves in the face of a fluffy snack showed greater achievements later in life, has since been questioned. A replication of the study published in *Psychological Science* in 2018 found that the correlation between the ability to delay gratification in childhood and achievement later in life was "half the size" of the effect reported in Mischel's original work. Furthermore, once you controlled for education, family life, and early cognitive ability, the correlation between denying a marshmallow and later behavior dropped to a big fat zero. Yet the marshmallow test and its follow-ups (though admittedly never intended to be used this way) helped shape public school educational policies.

Stanley Milgram and his shock tests — using the same machine that Rosenhan used during some of his early studies before he arrived at Stanford — have also been challenged. Psychologist and author Gina Perry revealed in her book *Behind the Shock Machine* that Milgram and his cohorts coerced participants into delivering shocks, which shows that the conclusions of the study — that we all are susceptible to blindly following authority — may not be as cut-and-dry as the experiment alleged, though there have been many replications of his research (including a 2017 paper out of Poland where 72 out of 80 participants were willing to shock other innocent subjects at the highest level).

Among the hardest hit has been Philip Zimbardo, the architect of the famous prison study, which took place in Stanford's basement in 1971 while Rosenhan was working on "On Being Sane in Insane Places." Zimbardo and his researchers recruited students from a newspaper ad and assigned them roles as "inmates" or "guards." Guards abused inmates; inmates reacted as real prisoners. One even famously screamed, "I'm burning up inside . . . I want to get out! . . . I can't stand another night! I just can't take it anymore!" The whole demonstration evidently revealed the ingrained sadism at the core of all of us, if given the power and opportunity. Zimbardo became an overnight expert and his work was even consulted in a 2004 congressional hearing on Abu Ghraib prisoner torture. When Zimbardo first saw the photographs of the abused, he told the *New York Times,* "I was shocked. But not surprised . . . What particularly bothered me was that the Pentagon blamed the whole thing on a 'few bad apples.' I knew from our experiment, if you put good apples into a bad situation, you'll get bad apples." Some argue that this perspective released the aggressors from responsibility. If we all have a monster inside of us, waiting to emerge in the right context, then how can we blame or punish people when it inevitably does?

The study, some say, even helped push the dial *away* from prison reform, since prison was deemed, thanks in part to Zimbardo, as "not reformable." But the study's critics, of which there were many, landed a few more concrete hits in recent years. A 2018 *Medium* piece by journalist Ben Blum blew up the internet (in certain circles). Blum had tracked down one of the "inmates" — the one who screamed "I'm burning up inside" — and found out that his pain was a performance. "It was just a job. If you listen to the tape, you can hear it in my voice: I have a great job. I get to yell and scream and act all hysterical. I get to act like a prisoner. I was being a good employee. It was a great time." Blum further discovered that Zimbardo had coached the guards and even thanked one of the more aggressive ones. "We must stop celebrating this work," personality psychologist Simine Vazire tweeted. "It's anti-scientific. Get it out of textbooks."

Psychologist Peter Gray, who had removed Zimbardo from his *Psychology* textbook in 1991, long before the *Medium* article, told me in an interview that he sees this as a "prime example of a study that fits our biases . . . There is a kind of desire to expose the problems of society, but in the process cut corners or even make up data." He said this is happening more often now because there are greater numbers of post-doctorates competing for fewer jobs and grant resources. "There is an epidemic of fraud."

This epidemic is not limited to social psychology, but is mirrored across all disciplines, from the heavily data-oriented fields of cancer studies and genetics to dentistry and primate studies. In 2016, Australian researcher Caroline Barwood and colleague Bruce Murdoch were convicted of cooking the books on a "breakthrough" study on Parkinson's — and nearly went to jail for it.

Korean stem-cell researcher Hwang Woo Suk and Harvard evolutionary biologist Marc Hauser are just two more celebrated academics to face allegations that they had fabricated their work and committed academic fraud. This of course happens when there is big business interest outside of academia, too. There's Elizabeth Holmes and her blood testing company, Theranos, which raised $700 million before the *Wall Street Journal*'s John Carreyrou helped expose the company to be a "massive fraud." Richard Horton, editor of the *Lancet,* wrote in a 2015 op-ed, "Much of the scientific literature, perhaps half, may simply be untrue . . . Science has taken a turn towards darkness." One of the leaders of the push to uncover academic fraud is Stanford's John Ioannidis, who authored a scathing 2005 paper titled "Why Most Published Research Findings Are False." He's found that out of thousands of early papers on genomics, only a tiny fraction stood the test of time. He then followed forty-nine studies that had been cited at least a thousand times, and found that seven had been "flatly contradicted" by further research.

I notice fraud everywhere now. In the fall of 2018, Cornell University professor Brian Wansink resigned after thirteen of his papers — including one that showed how serving bowl size affects food consumption — were retracted and Cornell found he committed "academic misconduct in his research and scholarship, including misreporting research data." That same time, thirty-one papers published by Dr. Piero Anversa, a former Harvard Medical School professor and cardiac stem cell researcher, were singled out as including "falsified and/or fabricated data" and retracted. If you want to see the scourge this has become on the field in real time, check out a blog called *Retraction Watch,* which strives to post every single academic retraction and keeps a top ten list of the most highly cited retracted papers.

And this fraud, played out every day in our academic journals and our newspapers (or more likely our social media feeds), breeds an anti-science backlash born of distrust. We've seen this most dangerously in the recent measles outbreak spurred on by the anti-vaxxer movement (whose theories are based around the fraudulent Wakefield study, published in the *Lancet,* one of the world's oldest and respected journals, and since retracted). How many times, people wonder, can we be told that this or that was "proven" in studies — only to be warned that the opposite is true the very next day — before we start to doubt all of it?

As we've seen, this doubt is particularly corrosive to psychiatry.

We still don't know exactly how so many psychiatric drugs work or why they don't work for a significant percentage of people. All of the current treatments for mental illness are "palliative, none are even proposed as cures." We still don't have clear-cut preventive measures and we still haven't figured out how to improve clinical outcomes for everyone or even how to prolong life expectancy. Though serious mental illnesses, like schizophrenia, clearly have heritable components, genetic research has yielded interesting but mostly inconclusive results.

The lay public today is fully aware of the deep connections between Big Pharma and psychiatry, which were cemented during the creation of the *DSM-III* and have only expanded since. No wonder there's been a fallout with the drugs, as direct-to-consumer advertising promised all manners of advancements and cures. But the new drugs, called "atypicals" or "second-generation" antipsychotics because they were marketed to have fewer side effects, have failed to deliver on many of their promises. Second-generation drugs come with their own issues, including excessive weight gain and metabolic disorders, and, in 2010, the *New York Times* reported that they were "the single biggest target" of the False Claims Act, resulting in billions of dollars spent settling charges of fraud. (Johnson & Johnson, for example, agreed to pay $2.2 billion in 2013 for hiding the host of side effects of its drug Risperdal, which include stroke and diabetes.)

Author and journalist Robert Whitaker, who has created a powerful arena for challenging traditional psychiatry on his blog, *Mad in America,* based on his 2001 book of the same name, sums up the outrage: "For the past twenty-five years, the psychiatric establishment has told us a false story. It told us that schizophrenia, depression, and bipolar illness are known to be brain diseases . . . It told us that psychiatric medications fix chemical imbalances in the brain, even though decades of research failed to find this to be so. It told us that Prozac and the other second-generation medications were much better and safer than the first-generation drugs, even though the clinical studies had shown no such thing. Most important of all, the psychiatric establishment failed to tell us that the drugs worsen long-term outcomes."

In the face of such rampant distrust, some of the "best and brightest" cling to their arsenal with a delusional level of certainty. A well-known psychiatrist (whom I will allow to remain nameless since he doesn't see patients these days — he's that high up; apparently the more successful you are, the fewer hours you spend with patients) lectured me about how to fix the broken system: "They just need to take their drugs," he said, sipping his wine. "What we have is just as effective as the drugs that treated you." The blind arrogance of this comment made me laugh out loud. Though some people are bigger drug advocates than others, most reasonable doctors acknowledge the limitations of psychiatric medications. The hardest part of coping with a serious mental illness, according to people I've interviewed who live with one, is the more subtle negative symptoms — the cognitive impairments, the parts of the illness that make life harder to navigate and are not ameliorated by the medications available. It feels like "your life is taken away from you. That all the things you once enjoyed are gone," said a twenty-year-old who had recently been diagnosed with schizophrenia.

But I'm not here to rail against the drugs. There are plenty of places where you can get that perspective. I see that these drugs help many people lead full and meaningful lives. It would be folly to discount their worth. We also can't deny that the situation is complicated. If I know this and you know this, then that arrogant doctor, a leader in the field, does, too. Yet there he sits, sipping wine and spouting absurdities.

The reputation, the distrust, the lack of progress: All of this has contributed to a worldwide shortage of mental health care workers. Some say it's the pay — for many years psychiatrists were the third lowest paid medical specialists (though this, as we'll see, is starting to change). Psychiatry was once seen as a humanistic medical science; as of 2006 only 3 percent of Americans receive any kind of psychotherapy — from "problem based" cognitive behavioral therapy to open-ended psychodynamic treatment. Freud's officially "dead." His work has been reexamined as "sexist, fraudulent, unscientific, or just plain wrong . . . Psychoanalysis belongs with the discarded practices like leeching." In the meantime, psychiatry has shifted from a soft science to a hard one, and in doing so has become largely mechanical and mundane.

These issues partially explain why I didn't receive the smug reaction from the mental health community that I expected when I started to share my investigation outside the small world around Rosenhan. A few expressed shock, but many claimed not to be surprised. Psychiatrist Allen Frances listened to my case, then interrupted: "Before we get to that, could you go after the Koch brothers next?" But then he let the news sink in. That study was key to Robert Spitzer's work. Without it, "Spitzer could never have done what he did with *DSM-III*," he said. To find out that at least part of it was flimsy — if not worse — was far from vindicating; it was disheartening.

One psychiatrist friend started ranting about how the study was "ridiculous" and that Rosenhan's focus on labeling was "total bullshit." She wouldn't concede that his larger points — namely about how patients are treated *because* of those labels — had any validity. Eventually she got so red in the face that I promised I wouldn't bring it up again.

At a research conference in Europe, where I was invited to speak about my illness, I agreed to meet a small group of research-oriented psychologists and psychiatrists for dinner after my talk. We met at a hotel bar that seemed plucked out of Midtown Manhattan, and joined four people at a table, all of whom were drinking martinis. I ordered a Manhattan, ignoring a voice warning me that it was never a good idea to drink bourbon cocktails at a professional event with strangers. The psychiatrists joked that they were going to "stay on New York time" so that they could just party through the conference. They talked a bit about my presentation and asked a few questions, but it was clear they were in vacation mode so the questions veered off-track.

One person asked: "How do schizophrenics feel about your book?"

I wasn't aware that there was one way that people with schizophrenia felt about anything, let alone my book. I looked back at him blankly, until one of the psychologists spoke for me. "Schizophrenics don't read." No one reacted. Was this a joke or was this truly the way a clinician felt about his patients?

Later, at a crowded restaurant, our table got rowdier the more alcohol we consumed. At some point, the subject of Rosenhan came up and I spoke a little about my research.

The psychologist who'd made the comment about people with schizophrenia not reading interrupted me. "I don't understand why you're even focusing on this study," he said, his voice thick. "I have no idea why you would do something that is so anti-psychiatry."

When I told him about my growing suspicions about the study, he got even more aggressive.

"Something like this is bad for all of us," he said, making a sweeping motion around the table, his voice rising in the now near-empty restaurant. The same person who was happy to dismiss the study as "anti-psychiatry" immediately raised his hackles at evidence that it wasn't aboveboard. Could it be that keeping this study solid benefited the narrative sold to many people in and outside the field — that we're making steady progress, that the bad old days are behind us?

"You have an opportunity to do something good and instead you focus on *this,*" he said, now pounding on the table. "Whether you like it or not, you're a symbol, and you should do something good with that power."

Perhaps it was the jet lag, or the latent frustration of getting nowhere with the pseudopatients, or the growing certainty that the study was fabricated and the feelings of disappointment I had about the man

481

behind it, or the mixture of red wine and Manhattans. Perhaps it was the fact that he called me a symbol (a symbol of what?). Whatever the cause, I lost it. I disappeared into the restaurant's closet-size bathroom, gazed into my own bleary eyes in the mirror, and mouthed, *Get yourself together* — remembering my own mirror image, the one who would not thrive as I had. I calmed myself enough to return to the table, my eyes red and my mascara smeared, where I couldn't help but launch back in. "I'm not trying to attack psychiatry. Give me a positive story to write and I will," I said, standing at the head of the table and speaking too loudly.

He looked up at me, resigned, put down his wine, and said, "Give me ten years."

We don't have ten years.

27
MOONS OF JUPITER

Taunted by death, chilled by the unknown, reproached by ambiguity, we doctors defy the dark, brandishing whatever truthiness we might have at our disposal. Humours, meridians, alchemy, or molecular biology, our scientific beliefs themselves are not as important as is the slim and ultimately betraying comfort they temporarily provide.
— Rita Charon and Peter Wyer,
"The Art of Medicine," *Lancet*

I don't know what happened to the young woman — my mirror image — who was misdiagnosed with schizophrenia for years before finally getting the proper diagnosis. Once she left the psychiatric hospital, the doctors lost track of her and she became just another patient with a poor prognosis, once an interesting case, now another name in the files. Did she surpass her doctors' low expectations and surprise everyone with a miraculous recovery, as I did? Or is she simply another casualty of bad timing?

For every miracle like me, there are a hundred like my mirror image; a thousand rotting away in jails or abandoned on the streets for the sin of being mentally ill; a million told that it's all in their heads. As if our brains aren't inside those heads, as if that warrants dismissal, not further investigation. As if there could be any other response but humility in the face of the devastating enigma that is the brain.

"I think we should be honest about — acknowledge how limited our understanding is," Oxford psychiatrist Belinda Lennox told me. "That's the only way we'll do better."

Being honest about our limitations, as Dr. Lennox suggests, involves taking a harsh look at our history and the "truths" that we've accepted at face value. If solutions seem too good to be true, too categorical, too concrete, they usually are. When nuance is lost, medicine suffers.

That's where David Rosenhan and his paper come in. Rosenhan's study, though only a sliver of the pie, fed into our worst instincts: For psychiatry, it bred embarrassment, which forced the embattled field to double down on certainty where none existed, misdirecting years of research, treatment, and care. For the rest of us, it gave us a narrative that sounded good, but had appalling effects on the day-to-day lives of people living with serious mental illness.

Rosenhan did not create these outcomes, but his study enabled them. And now psychiatry is overdue for a reassessment of the terms we deploy, the new technologies on the horizon, the way we treat the sickest.

The psychiatric community, and society at large, is finally starting to rethink our terminology, which drives our social and health policies. Some, like advocate DJ Jaffe, argue that the mental illness net is far too wide and that we should focus on the 4 percent of the population who are most seriously ill, devoting the bulk of our funds to their treatment, instead of to the "worried well," whom psychoanalysts catered to in Rosenhan's era.

On the other side of the aisle, Dutch psychiatrist Jim van Os, who wrote "The Slow Death of the Concept of Schizophrenia and the Painful Birth of the Psychosis Spectrum" in 2017, believes we should put mental illness on a continuum. The big fat manual that is the *DSM* should be condensed to "not more than ten diagnoses," Dr. van Os told me, umbrella terms, like *psychosis syndrome* and *anxiety syndrome,* with gradients of symptoms, he argues. Dr. van Os believes this is the honest approach: It concedes, *Hey, we really don't know.*

The research community has reached similar crossroads. "Is schizophrenia disappearing?" one academic article asks; another poses the question: "Should the label schizophrenia be abandoned?"

There already are real-world implications to these queries. During his tenure as the director of the NIMH — the second longest ever in the agency's history — Dr. Thomas Insel implemented a new system that eschews *DSM* criteria called the *Research Domain Criteria.* The *RDoC,* as it's called, breaks down clunky labels like schizophrenia into their component parts: psychosis, delusions, memory impairment, and so on, rendering the wide concept of schizophrenia as scientifically meaningless in a research setting. (Insel has since left the NIMH for greener Silicon Valley pastures, and his *RDoC* has not been universally accepted — half of NIMH-funded studies still rely on *DSM* diagnoses. At this point the *DSM,* it seems, is too widely entrenched in the field to be fully replaced.)

Now instead of seeing something like schizophrenia as a monolithic entity —

almost too massive to study — people want to approach the disorder the same way we do cancer, by acknowledging the unique qualities of every case. The sheer variety of what we call schizophrenia might alarm anyone who does not personally know someone with schizophrenia. Some exhibit robust psychosis with delusions and paranoia, some hear voices, some have greater cognitive impairments and are more socially isolated, some are professors, some forgo hygiene, some become hyper-religious, some lose a great deal of their memories, some navigate the world without appearing to have any symptoms, and others don't speak at all and sit in a catatonic stupor. Some respond to drugs and live full and meaningful lives; some — from 10 to 30 percent — recover; others never do. But we don't hear about the variety. Instead, we get people like the psychiatrist in London asking me how schizophrenics feel about my book. Instead, we see the most extreme cases, the ones who end up on the streets with chronic forms of untreated psychosis. And so the narrative goes: Once you've been touched, you're lost.

What is now almost gospel is that the umbrella terms we use, like schizophrenia, have many causes, and that we should use "the schizophrenias" or "psychosis spectrum disorders," which gestures to the scant consensus about etiology. This perspective is partially due to genetic studies on serious mental illness, which have thus far remained inconclusive. Genetics is such a challenging area because there is not one gene associated with each disorder (as is the case with cystic fibrosis, which involves a mutation of one specific gene), but hundreds. However, several studies have now revealed a "genetic overlap" in psychiatric disorders, especially among bipolar disorder, schizophrenia, major depressive disorder, and attention-deficit/hyperactivity disorders. "The tradition of drawing these sharp lines when patients are diagnosed probably doesn't follow the reality, where mechanisms in the brain might cause overlapping symptoms," said Ben Neale, an associate professor in the Analytic and Translational Unit at Massachusetts General Hospital. This may just offer scientific proof for what many in and outside the field have been saying for so long: The hardline differences among the terms we use do not have scientific validity.

It's telling that the more we open our eyes to what we don't know, the more excitement builds in the research community. Emerging studies exploring the link between the immune system and the brain — as is the case for autoimmune encephalitis — have galvanized the quest to understand how thoroughly the body itself influences and alters behavior, spurring studies of immune-suppressing drugs on people with serious mental illnesses. Researchers have estimated that as many as a third of people with schizophrenia display some immune dysfunction, though what that means about the underlying cause of the illness remains unclear.

An interest in the connection between the gut and the brain has led to some fascinating research on probiotics, which have been shown to reduce mania and some of the more robust symptoms of schizophrenia. Psychiatric epidemiologists are also finding that people born in winter months — during times of heightened flu and viral infections — may be more likely to develop serious mental illness (though people with more *severe* forms of the illnesses are more likely to be born in the summer months, so who knows). There are examples of psychosis brought on by gluten intolerance or cured by a bone-marrow transplant; of people misdiagnosed with serious mental illness who had Lyme disease or lupus. The more we learn about the body and its interaction with the brain, the more the shroud begins to lift.

Meanwhile, new technologies are also providing deeper access to the brain than ever before. "What I teach my students is, 'How did Galileo manage to demonstrate the veracity of the Copernican view of the [sun-centered] universe?' Well, the main advances were incremental in their ability to refine glass into lenses. Not very sexy, except that he could use that to make his own telescope and see the moons of Jupiter," Dr. Steven Hyman of the Broad Institute told me. Hyman was admittedly "giddy" after his institute published a highly touted paper in *Nature* in 2016 that linked schizophrenia with a protein called complement component 4 (C4), which plays a role in "pruning" the brain in young adulthood, marking unnecessary synapses that should be removed as the maturing brain hones itself. Though only in its early stages, this line of inquiry provides a model of schizophrenia that might involve such "overpruning."

Greater tools are on the horizon (or already here) to allow us to peer into the still-mysterious machinations of the brain, including Drop-Seq, which one day may provide the cell-by-cell census of the brain;

optogenetics, which manipulates brain circuits in live animals using light; CLARITY, which melts away the superstructure of the brain, making tissue transparent as a way to look at the fine structure of cells three-dimensionally; and a new technique (described in *Science*'s January 2019 issue) that uses 3-D technology and higher resolution to pinpoint individual neurons in record time. Labs across the country are also making stem cells out of skin cells from people diagnosed with mental illnesses and manipulating them in order to understand how the brain functions or malfunctions. They are in essence creating "mini-brains" (at this very moment!), which will allow them to study in real time how medications affect each individual brain.

IBM's Watson team told me about plans to create "Freud in a box." Their hope is to get Watson qualified as a psychiatrist. Watson would not replace psychiatrists, they explained; to the contrary, the computer algorithm would give psychiatrists more time to really talk to the patient and interact human-to-human. Some psychiatrists tell me that they are enthusiastic about wearable tech, which would give them access to mountains of data that were formerly self-reported. "Digital phenotyping" could chart everything from how active a person is to how often she opens the fridge to how many times a day she logs in to her social media accounts. Passive listening devices could monitor the content and tone of speech. There are wearable "galvanic" skin sensors that could create biofeedback on anxiety levels. There are even swallowable sensors that could tell a doctor if you're taking your meds, as well as ongoing studies using virtual reality programs as treatment for phobias. As exciting (and, yes, ominously Big Brother) as this sounds, it doesn't get us closer to fixing the validity issues at the core of diagnosis. Data alone will not give us the answer to the question: *If sanity and insanity exist, how shall we know them?* But it might help.

This new enthusiasm is starting to breed a new faith. Or, at least, it looks that way (I've learned to be wary of putative easy fixes). The old guard tells me that they have begun to notice something that has long been missing: optimism. More medical students are pursuing careers in the field, and, perhaps not coincidentally, after years of modest gains, the average psychiatrist's salary increased more than that of any other specialty in 2018 — higher than the take-home pay of immunologists and neurologists. "We have never seen demand for psychiatrists this high in our 30-year history," a physician recruiting firm said in 2018. "Demand for mental health services has exploded."

Also promising are the indicators that the damning distrust created by years of psychiatry's coziness with Big Pharma has started to self-correct. While psychiatry has become more transparent about its connections, pharmaceutical companies have begun devoting less funding to psychiatric research — decreasing its flow to those areas by 70 percent over the past decade after so many drugs failed to beat placebos, or after the expiration of lucrative patents (Zyprexa, Cymbalta, Prozac, to name a recent few). Though lost research dollars never sounds like a good thing (and a loss of investment in finding new advancements is certainly not), a few smaller, niche companies are stepping in and focusing on psychiatric research — looking to investigate new drug pathways and incorporating genetics into treatment (a field called pharmacogenetics). "It is to be hoped that a younger generation of researchers will break out of the confines of traditional theorizing that started a process but left the path to its conclusion obscure," wrote veteran researcher psychiatrists Dr. Eve Johnstone and Dr. David Cunningham Owens in *Brain and Neuroscience Advances* in 2018. In other words, fresh eyes may just open up a new path.

And as it turns out, the advances in pharmacology don't even need to be new. Another exciting path happens to have been paved long ago. After years of being stymied by the War on Drugs, which made research of Schedule 1 drugs almost impossible, we are now in the midst of a psychedelic revival. Clinicians now use LSD and psilocybin as treatment for everything from depression to PTSD. Even brain stimulation, which originated in the 1950s as a way to "treat" homosexuality and schizophrenia, is making a comeback of sorts. Some techniques involve implanting electrodes that send electric pulses right to specific brain tissue, another attaches the electrodes noninvasively on the scalp. These procedures are now gaining ground in top hospitals for treatment of a host of issues from OCD to depression to Parkinson's. Meanwhile, a variation of the anesthetic ketamine (developed in 1962 and nicknamed "special K" by club kids in the '80s and '90s) was recently approved by the FDA for use in treatment-resistant depression, which affects 20 percent or more of people with the disorder. It is quite striking to see a drug that's been around since Rosenhan's era being touted on all the morning shows as one of the biggest breakthroughs in psychiatric medicine in the last fifty years.

And then, after years of being dismissed as a soft science, talk therapy, too, has seen a reconsideration as studies have shown that for some people, therapy creates profound changes in the brain — as pronounced, in some cases, as psychiatric medication. "Psychotherapy is a biological treatment, a brain therapy," said Nobel Prize–winning psychiatrist and neuroscientist Eric Kandel in 2013. "It produces lasting, detectable physical changes in our brain."

"One sees as far as one is limited by the technologies of that time," said *Lancet Psychiatry* editor Niall Boyce. "If we were to draw an analogy, I'd say that we're at the point of infectious disease research [when] the microscope has just been invented, and the story is really sort of getting cracking." Child psychiatrist and geneticist Matthew State of the University of California at San Francisco uses a similar analogy, adding: "It's true, [it's like having] a microscope for the first time. And it's not like a single microscope; we have like three different microscopes that we've never had before."

There is, some say, a lot to be encouraged about.

Even Dr. Torrey, the same man who told me that the field hasn't truly advanced since 1973, is optimistic. "You're going to see the whole thing turn around," he said.

"You think so?" I asked.

"Oh yeah. Keep your notes. Thirty to forty years from now, you'll be writing something totally different."

But we can't sit back, fold our hands, and wait for the future to solve all our problems, because even if we gain all the insights we seek, there is still the unresolved issue of basic care and treatment on the ground level. So while this or that brain imaging technology makes its way through the ivory towers of academia, people are still languishing on the streets, hidden in the general population, or behind bars, neglected by all of us.

In response to this tragic situation, people like UCLA psychiatrist Joel Braslow, a practicing psychiatrist and historian, have come to this position: "In spite of the fact that state hospitals were incredibly over-crowded and often being seen as custodial . . . at least we were caring for people then. Now we're not."

The late neurologist Oliver Sacks agreed, writing in his essay "The Lost Virtues of the Asylum" that "we forgot the benign aspects of asylums, or perhaps we felt we could no longer afford to pay for them: the spaciousness and sense of community, the place for work and play, and for gradual learning of social and vocational skills — a safe haven that state hospitals were well-equipped to provide."

When I first heard this reexamination, I thought of Nellie Bly. *We've already tried that, and see how well it worked out?* The snake pits of the past were a grotesque chapter in the history of medicine and the last thing we want to do is return to them. Yet you can't call what's happening now progress.

Three University of Pennsylvania ethicists, Dominic Sisti, Andrea Segal, and Ezekiel Emanuel, wrote an unfairly pilloried article subtitled "Bring Back the Asylum" in 2015. In it they argue persuasively for a new model of care that takes the best of the past and adapts it to a modern medical setting. No one can improve without the bare minimum — shelter, clothing, and food — but they also need care: intelligent medical intervention, personal contact, community, and meaning. In a perfect world (where money flowed freely into medical care), the authors envision a comprehensive approach that could provide all of the above — from full-time, inpatient care for the most acute and long-term beds for the most chronic, to community-based, family-supported outpatient therapy for those recovering: a tiered system (with ICUs, step-down units, and rehabilitation centers) like those that treat people with nonpsychiatric maladies.

Still, the paper's authors faced rampant criticism when their article was published — and Dominic Sisti even lost a contract with the city of Philadelphia's Division of Community Behavioral Health. One of the people who decided to pull his funding called his work "a disgrace."

"The debate boils down to one question: What counts as a mental disorder?" Dominic said. "Fights about involuntary treatment, long-term care: If you dig down deep enough, it comes down to the disagreement over the fundamental concept of a 'mental disorder.' These are the stakes."

These impossible questions have plagued us forever — the physical versus mental, brain versus mind — and they have profound, life-or-death consequences. As time marches on, the goalposts may move and the definitions may change, but it's the same story again and again — we consider some illnesses to be worthier of our compassion than others. And this has to change.

The change doesn't just require adding beds somewhere and letting people languish, it means taking a broader look at the infrastructure of each person's life — past and present — and the myriad ways environment shapes both sickness and health.

"The brain is extremely plastic," Dr. Maree Webster, the director of the Stanley Medical Research Institute's Brain Research Laboratory, told me. "Every experience you have changes your brain in a certain way. And so all this stuff — I mean it's very out of fashion thanks to psychoanalysis — but [early-life experiences], parenting, child abuse, these things increase your risk of developing a mental disorder." Environmental factors — obstetric complications, living in an urban area, experiencing trauma as a child, migrating to a new country, using cannabis, even owning a cat[1] — may in-

1. Some have suggested that people with schizophrenia are more likely to have antibodies directed against a common feline parasite (*Toxoplasmosis gondii*) that can also infect humans. Schizophrenia, studies say, is more common in countries where people keep cats.

crease the risk of developing a serious mental illness. In the United Kingdom, for example, the higher incidence of schizophrenia found in the Caribbean population has been linked to social factors like migration, social isolation, and discrimination.

Living in cities is linked to higher rates of schizophrenia. Why? It's unclear, but many have posited that urban environments are missing one element found in smaller, more tight-knit areas: support and community, a key part of the healing element found on Ward 11 and in Harry Lando's hospitalization.

Other research bolsters this conclusion. A two-year government-funded study published in the *American Journal of Psychiatry* has shown that early intervention after "first breaks" — or the first time experiencing the profound symptoms of serious mental illness — involving antipsychotic medication management combined with a "comprehensive, multi-element approach," which includes family support and psychotherapy, created the best outcomes.

New research and treatment models have emerged to train people who are troubled by hearing voices to better manage their lives alongside the auditory hallucinations — not by shutting the voices out completely but by interacting with them directly. Yale researchers found that a key difference in the hallucinatory experiences of psychics and of people with schizophrenia was that the psychics placed the voices in the context of a spiritual or religious experience, and were less disturbed by them. These new approaches to treating voice-hearing are supported by researchers at Stanford, who compared the experience of auditory hallucinations in people diagnosed with schizophrenia in the United States with those in developing countries. In America, where patients tend to subscribe to the biological model of mental illness, the patients reported antagonistic relationships with their hallucinations; the voices themselves were more likely to include violent, aggressive, and negative content. People in Chennai, India, and Accra, Ghana, by contrast, described more positive communions with their voices and reported better long-term outcomes. "Are those cultural judgments the cause of the illness?" Stanford anthropologist Tanya Marie Luhrmann asked. "Absolutely not. Do those cultural judgments make it worse? Probably."

One popular therapy that takes these cultural judgments into account is open dialogue therapy, which is meant to create an immersive community support system that practitioners say eventually allows for the reduction of antipsychotics while mining the content of a person's psychotic experience (which sounds like it would happily coexist with therapy at Soteria House or at R. D. Laing's Kingsley Hall). Open dialogue has emigrated from its birthplace in Finland to McLean Hospital in Massachusetts, a private psychiatric hospital ranked number one in the US. I saw McLean's version of open dialogue therapy in person and was struck by how simple it was: They treated the patient like a person.

The best in the field are expert at doing exactly that — relating to people and identifying symptoms that may be too subtle to pick up with other, more objective medical measures. This requires long meetings with patients where in-depth histories are taken and a sense of trust develops. Psychiatry at its best is what all medicine needs more of — humanity, art, listening, and empathy — but at its worst it is driven by fear, judgment, and hubris. In the end, the takeaway, repeated again and again in my interviews, is: Medicine in general, and psychiatry in particular, is as mysterious and soulful as it is scientific.

■ ■ ■ ■

You've heard of the placebo effect — its press is almost as bad as psychiatry's. The term originated in a religious context with the psalm Placebo Domine, "I shall please the Lord," but by the fourteenth century had taken on a more negative association within the church to describe fake mourners paid to attend funerals to "sing placebos" about the dead. The word made its way into medicine five centuries later, when, in 1772, Scottish physician and chemist William Cullen gave his patients mustard powder treatments for all manner of ailments, even though he knew it was a sham: "What I call a placebo." After World War II, researchers started to use sugar pills as controls to gauge the effects of "real" medicines. By the 1960s, the FDA had set double-blind placebo-controlled studies as the gold standard. And over time these seemingly inert sugar pills were found to have *measurably physical* effects on the body — though these effects were often viewed as illegitimate, mere noise that often got in the way of drug approval. Now we know that placebos set off complex parades of neurotransmitters — endorphins, dopamine, endocannabinoids, and others. If you receive a saline solution that you believe is morphine, your body reacts as if you have received six to eight milligrams of the drug — the equivalent of a pain-reducing dose. Parkinson's patients will release dopamine, sometimes even enough to control their involuntary movements, when they believe they're getting real L-Dopa drug treatments.

You can even augment the effects of a placebo with a caring, supportive environment, where the patient not only believes in the drug, but also believes in the doctor. Dr. Ted Kaptchuk, who heads the Program in Placebo Studies & Therapeutic Encounter at Harvard, pushes doctors to harness the power of the placebo in a more straightforward way. "Ultimately it's about being immersed in a world where we know that we're taken care of by healers, and that's the bottom line," Dr. Kaptchuk told me. "Every word counts, every glance counts, every touch counts. The five milligrams of a good medicine is very important, but it's more effective if you take it in the context of being aware that the healer, the doctor, the nurse, the physical therapist, also have an effect on the patient."

Just spending more time with the patient

can improve outcomes. In a study of acid reflux sufferers, a group that had a forty-two-minute consultation with a doctor did twice as well as one that had only eighteen minutes. To reflect their very *real* role in healing, some doctors are pushing to re-brand the placebo effect to "contextual healing," "expectation effects," or even "empathy responses."

This makes me think of my doctor Souhel Najjar, who had access to the fanciest, most cutting-edge tests, but whose breakthrough in my care came when he sat down on my bed, looked into my eyes, and said, "I'm going to do everything I can for you." My family and I believed in him; and I know in my core that his warmth and optimism helped me heal.

This faith in medicine, our healers, our diagnoses, our institutions is what Rosen-han helped shatter, what Spitzer helped rectify, and what the controversies over the *DSM-5* and the horror stories of our jail and prison systems have further shaken. Belief is what psychiatry has lost — and what it needs to survive.

This belief in a better way is what com-pelled the father at the beginning of this story to write to me about his son with schizophrenia. "Each time they tell me that schizophrenia is a lifelong condition, I ask them, 'What allowed Susannah Cahalan to escape the same diagnosis?' " he wrote in a follow-up email. Even as his son's condition has worsened, he has held on to a conviction that some change will come. I admire that so much.

This hope is essential. One mother told me about her experience navigating the mental health industry with her son, who was diagnosed with schizophrenia. After he started hearing voices in his teens, he was offered only a long list of medications that seemed to harm him more than help, because mainstream medicine insisted that *there was no cure for schizophrenia.* "If I'd adopted the conventional wisdom and accepted that my son would never get well, I'd have abandoned all hope," she explained. Instead, she tried everything else: orthomolecular treatment, which involved giving her son high levels of vitamin B, and trying out energy medicine, magnets, and "gemstone caps" that pulsed energy into his body. She met with shamans and holistic psychiatrists; she channeled dead ancestors, dosed him with plant essences, tried to remove the copper from his body, and bought him a device that would shield him from "e-smog," or electromagnetic radiation. Some people who hear this list might think she had lost her own grip on reality. But I don't think so. I think she is searching for options beyond merely surviving, searching for answers that might make her son happier and healthier. She continues to do so today. Can any of us blame her?

I refuse to plug my ears and continue to believe that we all live in a world where everyone finds their Dr. Najjar. I've met enough people like her in the trenches living with mental illness and talked to enough families advocating for ill loved ones to discount the disconnection between the dreams of the future and the reality of the present.

I know all too well that I am one of the lucky ones. My story is a bright and shining example of what can happen when cutting-edge neuroscience meets thoughtful doctoring in the most opportune conditions. More than piles of data or years of careful research, stories make us believe. And belief is a pedestal on which great medicine stands.

Even though I know it's a luxury many of us cannot afford, I still choose to believe. Though I'm painfully aware of terrible leads we've followed and the false promises of the past that were spurred on by bad science and blind hubris, I am still optimistic.

Yes, I'm skeptical when I hear of a new treatment or study "proving" this or that breakthrough, but I hold tight to the belief that what happened to me — finding a cure for what seemed to be "in the mind" — can happen to everyone. I've seen it happen over the years as I've traveled the country speaking on this topic, while also hearing the many heartbreaking ways medicine has failed others.

I believe in all the excitement emerging from neuroscience. And I believe that we will unravel the mysteries of the mind. I believe that we will puzzle out the seemingly unsolvable. I also believe that the puzzle is too complex for the human mind to grasp.

I am aware of all of the arrogance, incompetence, and failure, but I still believe that psychiatry — and the whole of medicine — will one day be deserving of my faith.

I believe. I believe. I believe.

EPILOGUE

"Whenever the ratio of what is known to what needs to be known approaches zero," Rosenhan wrote, "we tend to invent 'knowledge' and assume that we understand more than we actually do. We seem unable to acknowledge that we simply don't know."

Unlike Rosenhan, I don't want to "invent knowledge" where it's lacking. The truth is that I simply don't know so much. I know that David Rosenhan exaggerated and fabricated parts of his own story, the results of which were introduced on one of the most exalted pedestals in academia. I know Rosenhan's flawed work had an effect on Robert Spitzer and the creation of the *DSM*. I know that the study had a wide influence, contributing to the shuttering of psychiatric hospitals. I know that at least one pseudopatient's experience supported Rosenhan's thesis — and I know that one did not. I don't know why he never finished his book, why he never published on the topic again, or how he would feel about this book. I can guess, but I cannot know.

I don't know what happened to the other six pseudopatients. Did they exist at all? I will admit that I still keep imagining all the different ways a pseudopatient could un-mask herself (maybe I'll walk down the street one day, and I'll feel a light tap on my shoulder, turn around, and there you are). Because in the end I believe that he *exposed something real.* Rosenhan's paper, as exaggerated, and even dishonest, as it was, touched on truth as it danced around it — the role of context in medicine; the dismissal of psychiatric conditions as less legitimate than physical ones; the depersonalization felt by the mentally ill "other"; the limitations of our diagnostic language. The messages were worthy; unfortunately, the messenger was not.

When I had unearthed everything I could possibly find, I met with Lee Ross, the Stanford psychologist who introduced me to Rosenhan, and with Florence, my Rosenhan whisperer — the two living people intellectually closest with Rosenhan and most responsible for my obsession with him — to share my findings. Lee wrestled with his reaction to the news that Rosenhan may have fabricated his work. We sat in his living room and picked apart the arguments. Florence shared her perspective: "I was surprised initially when Susannah suggested that, but I don't find it reprehensible," she said. "I know I should, and it's science, but knowing David, David had a certain prankster quality to him."

Florence has seen as much of Rosenhan's files as I have and has no doubt that Rosenhan made up a good bulk of the paper, but she is more forgiving about the liberties he took. She likened him to a novelist creating a scene. She did not see him as a villain — she loved the man — but more as a rascal who had successfully punked the world; or, as she put it, as a latter-day Till Eulenspiegel, a prankster evoked in many German fairy tales, who "plays practical jokes on his contemporaries, exposing vices at every turn, greed and folly, hypocrisy and foolishness.

"What I've come to think about with David and this whole thing is his twinkle," Florence said. "You could imagine him saying, *Well, if I* had *completed this study, it would have been exactly as I described it.*"

Florence's acceptance of the possibility that Rosenhan's work may not have been completely legitimate opened something

511

inside Lee. "There is a certain shadowy quality when you probe into David's work and life," Lee Ross said. "It's just that feeling of you can't quite pin things down. Things don't quite add up sometimes. And I think he . . . I don't want this to have more connotations than it does. There is a way in which he kind of led multiple lives. And by that I just mean I think he was a somewhat different person in somewhat different contexts." I couldn't help but smile a little — that was, after all, one takeaway from Rosenhan's paper: that we're never all one way, that insane people are never always crazy, nor are sane people always rational. Lee continued: "I would be surprised, not unbelieving, but very surprised and very unhappy to learn [that he lied]. It would make me even more feel that David was struggling for a place in the sun."

Though I had to wonder: Was he struggling not for but *with* his place in the sun?

Rosenhan, with his twinkle and shadowy quality, managed to expose truths — even if those truths contained problematic fictions — and created something that we still debate, pillory, celebrate, and investigate nearly half a century later. The study may have "proved" something that people *believed* was true, and, for better or worse, that was enough to change everything. Maybe it's as Chief Bromden says in *One Flew Over the Cuckoo's Nest:* "It's the truth even if it didn't happen."

There were no lines around the block to attend Rosenhan's funeral. No national newspaper covered his passing. The sparse attendance was partially due to how inured Rosenhan's community had grown to their grief. A series of tragedies hit the aging professor with such senseless brutality that people could not help but make comparisons to Job. It started with his daughter Nina's death in 1996 in a car accident, followed by Mollie's lethal lung cancer diagnosis, followed by Rosenhan's first stroke — a small TIA that would likely have gone unnoticed if Rosenhan had not insisted that he be checked out. Florence noticed a slight difference in her friend after that first scare. With a mind so nimble, he was good at hiding it, but there was a new hesitation, a few seconds of delay that had never existed before. Mollie died in her bed at home in 2000, around the time that Rosenhan suffered a massive stroke from which he would never recover. The stroke and the other illnesses that had befallen him had damaged his vocal cords so that the familiar baritone voice faded into silence. The man who took daily multi-mile walks around Stanford's Dish, the professor who made you feel truly *seen,* the warm, approachable raconteur, turned into a shell of himself. He lost his ability to walk and moved into a nursing home. The stalwarts — among them his friend and caretaker Linda Kurtz; his son, Jack; and Florence — came frequently to check on him. Otherwise, people forgot about him. When I contacted his former friends and colleagues, many of whom had attended parties at his house for years, they inquired about how he was because they had not heard about his death.

At his funeral, Rosenhan's close friend Lee Shulman, who spent many hours studying the Talmud with Rosenhan in a study group, gave a speech that perfectly captured Rosenhan:

> David's fame was based on many accomplishments, but one stands out as a powerful beacon. His essay in *Science,* "On Being Sane in Insane Places," begins with an opening sentence that should be intoned in the register of the yeshiva student whom he would always remain: "If . . . sanity and insanity exist . . . how shall we know them?" . . .

If you have never actually read that article, or it's been a long while, you just may have forgotten its rhetorical power . . . It is a proclamation, a moral outcry, a scream of pain and a demand that the world bear witness.

David Rosenhan is no clearer a figure to me now, even after my years of relentless digging into his personal and professional past, than he was the day I first heard Dr. Deborah Levy talk about his study. He was, as Lee Ross said, "a somewhat different person in somewhat different contexts"; depending on what kind of light you cast on him, you could view him as a hero or a villain, a scoundrel or a rascal, a charlatan or a Cassandra, a selfless leader or a selfish opportunist.

But there's one story that sums him up for me, as a thinker, as a father, and as a human.

Jack was thirteen when his father invited him to join him on a trip to New York City to meet with an editor and discuss the pseudopatient book that he would never publish. The two were walking through the crowded streets of downtown Manhattan when they noticed an open grate on the sidewalk. Through the hole you could see below to a

515

whole hidden world. They nearly gasped when a huge dump truck drove by underfoot.

"Don't say a word, just follow me," Rosenhan said, leading his son to one of the hardhats manning an elevator that led belowground.

He introduced himself as David Rosenhan, professor of engineering at Stanford University. In a flash, Rosenhan and Jack were fitted for hard hats and boots. Zoom! They were in an elevator headed underground to see the building of the infrastructure of the New York City subway system firsthand. Their guide seemed impressed by Rosenhan and his credentials and gave them the full tour. Jack kept worrying that they'd be busted. *Just one complicated engineering question and we're toast,* he thought. But Rosenhan seemed as cool and confident as always, carrying himself as if he belonged there, as if he were the king of the underground, a world invisible to the people who walked in droves above them. This simple fact blew Jack's young mind: His father could so easily *become someone else.*

He was the great pretender.

ACKNOWLEDGMENTS

Five years ago, when I started researching this book, Dr. E. Fuller Torrey wrote me an email after our first meeting: "It is a good project for a layperson because you come to it with a fresh outlook, not contaminated by the revered wisdom of the professionals who may, but often do not, know what they are talking about." I like the sentiment (the email hangs above my desk) — and though I have come across many professionals who do not know what they're talking about, I've encountered many more who do. This is a list, by no means definitive, of the many generous people who took time out of their busy lives to help me write this book.

First and foremost, a heartfelt thank-you to Florence Keller and LaDoris Cordell, the Wonderful Women of Wilkie Way, who held my hand throughout years of research-ing and writing, providing support, wisdom, and counsel. You two have brought so much to my life and I'm forever grateful that David brought us all together. I could not have written this book without you.

This was not the book I initially intended to write. While researching it, however, I got to spend time with David Rosenhan's son, Jack, and his wife, Sheri, two supremely kind and generous people. I am grateful for your time — it is a joy to know you.

I so enjoyed the hours spent with Bill and Maryon Underwood. It was such a blast walking down memory lane. And to Harry Lando — the great footnote — thank you for being so open and honest. I hope I did your experiences justice.

It takes a small army to get what's on my laptop ready for public consumption. Thank you to agent duo Larry Weissman and Sascha Alper, who helped shepherd this project and find its perfect home. Thank you to the magnificent Millicent Bennett: You are a gift sent from writer heaven, and I cherish the day our paths crossed. Thank you for your tireless support, your brilliant mind, and your steadfast belief in this project. Thank you to Carmel Shaka for keeping us on track during a serious time crunch. Thank you to the dream team at Grand Central Publishing for championing this book, especially Michael Pietsch, Ben Sevier, fel-

518

low Hilltopper Brian McLendon, Karen Kosztolnyik, and Beth deGuzman. Thank you also to the powerhouse publicity team led by Matthew Ballast with help from Kamrun Nesa and Jimmy Franco and social media maven Alana Spendley. Thank you to the sales team — Ali Cutrone, Alison Lazarus, Chris Murphy, Karen Torres, Melissa Nicholas, and Rachel Hairston — for their early enthusiasm (even after I babbled at them with twin newborn "mom brain"). Thank you to the supertalented art and production team — Albert Tang, Kristen Lemire, Erin Cain, Carolyn Kurek, Laura Jorstad, and a special shout-out to Tareth Mitch, who, late one Friday, saved the day.

Thank you to my early readers: Dr. Dominic Sisti for his nuanced look at diagnosis and the role of institutions (as well as his support throughout the writing); Dr. Andrew Scull for helping me understand Rosenhan's place in history and his infectious enthusiasm for the research; Dr. Will Carpenter for his perspective on the biological side of psychiatry; Dr. Len Green for his perspective on the history of psychology and the replication crisis; Dr. Michael Meade for his general wisdom; Dr. Craig Haney for taking time to help me understand the range of horrors happening in jails and prisons. Thank you to Dr. Belinda Lennox, who read an early draft and urged me to be a bit softer on the field; and thank you to the brilliant Maureen Callahan, who pushed me to be a bit harder on it. Ada Calhoun and Karen Abbott, my beloved Sob Sisters, provided support and enthusiasm when I needed it most. Panio Gianopoulos, combination superman and mensch, helped me to control the chaos of the first drafts, and Karen Rinaldi helped me keep my head straight. Thank you to Dr. Niall Boyce for introducing me to the concept of a microhistory and to Allen Goldman for his unerring support and clarity during the book's final stages. Thank you to Hannah Green for her take on the complexities of the criminal justice system, and to Dr. Heather Croy, whose help with the twins made it possible for me to finish this book. Thanks also to Shannon Long and Emmett Berg for their help with research, and especially to the remarkable Glyn Peterson, who went above and beyond with her eagle-eyed fact-checking.

A special hat tip to Dr. Deborah Levy and Dr. Joseph Coyle for setting me off on this mission — who would have ever guessed that an offhand comment made in a crowded restaurant would consume the next five years of my life? Thank you also to Dr. Lee Ross, who helped stoke an early interest in Rosenhan and his famous study.

Thank you to the staff of Stanford University's Special Collections and Swarthmore College for letting me camp out and dig into the research. Haverford Hospital's lead researcher Margaret Schaus supplied me with a treasure trove of primary sources, as did the Historical Society of Pennsylvania. Thank you to the Treatment Advocacy Center, especially to E. Fuller Torrey and Maree Webster, who gave me a tour of the Stanley Medical Research Institute's brain bank. Thank you also to the Center of Inquiry for hosting a strange and fun research trip. Thank you also to Emilie David at *Science* for her help in tracking down documents, and thank you to DJ Jaffe for taking time out to school me on all the facts.

And to the staff and patients at the following hospitals, thank you for hosting me: McLean Hospital (especially Dr. Bruce Cohen, Dr. Dost Ongur, and Dr. Joseph Stoklosa), Santa Clara Valley Medical Center, Zucker Hillside Hospital and the staff of the Early Treatment Program, and University of Pennsylvania's PEACE program (especially Dr. Irene Hurford).

There are mind-blowing museums across

the country devoted to the history of psychiatry, but many are hidden away from the general public. Thank you to social worker and historian Dr. Anthony Ortega for his unforgettable tour of the Patton State Hospital Museum, and to Bethlem Hospital and the Institute of Living for allowing me to visit their collection.

Thank you to Dr. Michael First for being such a good sport; to Dr. Nancy Horn for her passionate perspective; to Dr. Janet Williams for bringing Dr. Robert Spitzer to life; to Mary Bartlett and Claudia Bushee for embracing my intrusive questions about your family. Thank you to Dr. Allen Frances for his perspective on the *DSM;* to Dr. Gary Greenberg and Dr. Ian Cummins for helping me figure this story out; to Drs. Ken and Mary Gergen for giving me my aha moment; to Dr. Karen Bartholomew for going above and beyond; thank you to Dr. Jeffrey Lieberman for the four-part history lesson; thank you to Dr. Matthew State and Dr. Steven Hyman for making me excited about the future of the field; thank you to Dr. Chris Frith and Dr. Thomas Insel for answering my many stupid questions with such patience. Thank you to IBM's Watson team (especially to Guillermo Cecchi) for inviting me to your headquarters. I am indebted to Ron Powers for his beautiful *No One Cares About Crazy People* and the perfectly inscribed copy of *Good Night Moon*. I'm so grateful to Justen Ahren and the Noepe community for giving me a place in the most beautiful corner of the earth to write.

For their perspectives on Rosenhan, the man, thank you to: Dr. Edith Gelles, Dr. Helena Grzegolowska-Klarkowska, Abbie Kurinsky, Linda Kurtz, Dr. Miv London, Vivian London, Pamela Lord, Harvey Shipley Miller, Dr. Kenneth P. Monteiro, Hank O'Karma, and Dr. Lee Shulman.

For their perspective on David Rosenhan, the psychologist, thank you to: Robert Bartels, Dr. Daryl Bem, Dr. Gordon Bower, Dr. Bruno Breitmeyer, Dr. Allen Calvin, Dr. Gerald Davison, Dr. Thomas Ehrlich, Dr. Phoebe Ellsworth, Drs. Raquel and Ruben Gur, Dr. Eleanor Maccoby, Dr. David Mantell, Bea Patterson, Dr. Henry O. Patterson, Dr. Robert Rosenthal, Dr. Peter Salovey, Dr. Barry Schwartz, Dr. Martin Seligman, Dr. Ervin Staub, and Dr. Philip Zimbardo.

For their perspectives on the study, thank you to: Dr. Matthew Gambino, Dr. Peter Gray, Dr. Benjamin Harris, Dr. Voyce Hendrix, Dr. Marc Kessler, Dr. Alma Menn, Dr. John Monahan, Dr. Gina Perry, and Dr.

Christopher Scribner.

For their perspective on psychiatry's past, present, and future, thank you to: Richard Adams, Dr. Justin Baker, Dr. Gary Belkin, Dr. Richard Bentall, Dr. Carol Bernstein, Claire Bien, Dr. Joel Braslow, Dr. Cheryl Corcoran, Dr. Philip Corlett, Dr. Anthony David, Dr. Lisa Dixon, Mark Gale, Dr. Steven Hatch, Dr. Robert Heinssen, Dr. John Kane, Dr. Ken Kendler, Dr. Richard Lamb, Dr. Robert McCullumsmith, Kerry Morrison, Dr. Souhel Najjar, Dr. Stephen Oxley, Dr. Roger Peele, Dr. Thomas Pollack, Dr. Steven Sharfstein, Dr. Kate Termini, Dr. Jim van Os, Dr. Mark Vonnegut, and Bethany Yeiser.

Thank you most of all to Stephen Grywalski. The past four years were intense — a marriage, a run-in with Marie Laveau, an ileus, a move — you are my tireless advocate and you've given me the greatest gift of all: our twins, Genevieve and Samuel. Without you, none of this.

NOTES

I relied on a treasure trove of materials to put together this book — most notably from Florence Keller's file of "On Being Sane in Insane Places"–related documents. Stanford Special Collections also provided eight banker's boxes' worth of documents from David Rosenhan's three-decade career. I relied on his diary entries, his unpublished book, audio and video recordings of his interviews and lectures, newspaper interviews, and television and radio appearances, and I interviewed hundreds of people who knew him. Research on the history of psychiatry came from a wide variety of sources, many listed here, including interviews with experts in the field, site visits to psychiatric hospitals, and archival research. Still, I've only scratched the surface of the history of mental health care. Take a look at notes below for references to other, more in-depth sources. And if the spirit moves you, read them.

Preface

Patient #5213's . . . Details like this one in the preface came from medical records found in David Rosenhan's private files.

"Do you recognize the voices?" . . . Direct quotes are from Rosenhan's unpublished book, *Odyssey into Lunacy,* chapter 3, 5–6.

"The history of psychiatry" . . . Edward Shorter, *A History of Psychiatry: From the Era of the Asylum to the Age of Prozac* (Hoboken, NJ: Wiley, 1996), ix.

Part One

Much Madness is divinest Sense . . . Emily Dickinson, *The Poems of Emily Dickinson* (Boston: Roberts Brothers, 1890), 24.

1: Mirror Image

"assess both the mental and physical" . . . American Psychiatric Association, "What Is Psychiatry?," https://www.psychiatry .org/patients-families/what-is-psychiatry.

"Psychiatry has a tough job" . . . Dr. Michael Meade, email to Susannah Cahalan, March 17, 2019.

called the great pretenders . . . For a discussion of these disorders, see Barbara Schildkrout, *Masquerading Symptoms: Uncovering Physical Illnesses That Present as Psychological Problems* (Hoboken, NJ: Wiley, 2014); and James Morrison, *When Psychological Problems Mask Medical Disorders: A Guide for Psychotherapists* (New York: Guilford Press, 2015).

"the lay public would be horrified" . . . Dr. Anthony David, phone interview, January 28, 2016.

the one in five adults . . . "Mental Illness," National Institute of Mental Health, https://www.nimh.nih.gov/health/statistics/mental-illness.shtml.

urgently affects the 4 percent . . . "Serious Mental Illness," National Institute of Mental Health, https://www.nimh.nih.gov/health/statistics/prevalence/serious-mental-illness-smi-among-us-adults.shtml/index.shtml.

"mental, behavioral or emotional disorder" . . . "Serious Mental Illness," National Institute of Mental Health.

whose lives are often shortened . . . World Health Organization, "Premature Death Among People with Severe Mental Disorders," https://www.who.int/mental_health/management/info_sheet.pdf.

"Insanity haunts the human imagination" . . . Andrew Scull, *Madness in Civilization* (Princeton: Princeton University Press, 2015), 10.

your blue may not be my blue . . . For more on the variability of color perception, see Natalie Wolchover, "Your Color Red Really Could Be My Blue," *Live Science,* June 29, 2014, https://www.livescience .com/21275-color-red-blue-scientists .html.

"medically unexplained" . . . For more on the so-called medically unexplained, see Suzanne O'Sullivan, *Is It All in Your Head?: True Stories of Imaginary Illness* (London: Vintage, 2015).

how everyday drugs like Tylenol work . . . Carolyn Y. Johnson, "One Big Myth About Medicine: We Know How Drugs Work," *Washington Post,* July 23, 2015, https:// www.washingtonpost.com/news/wonk/wp/ 2015/07/23/one-big-myth-about-medicine -we-know-how-drugs-work/?utm_term= .1537393b19b4.

what exactly happens in the brain during anesthesia . . . Susan Scutti, "History of Medicine: The Unknown Netherworld of Anesthesia," *Medical Daily,* March 5, 2015, https://www.medicaldaily.com/history-med

icine-unknown-netherworld-anesthsia-32
4652.

a condition like anosognosia . . . "What Is
Anosognosia?" WebMD, https://www
.webmd.com/schizophrenia/what-is
-anosognosia#1.

"They seem to blame my son" . . . The
father who wrote this email to me prefers
to maintain his privacy. Email to Susannah Cahalan, March 7, 2018.

2: Nellie Bly

To re-create Nellie's preparation and hospitalization, I relied on her own writing: *Ten
Days in a Mad-House* (New York: Ian L.
Munro, 1887), https://digital.library.upenn
.edu/women/bly/madhouse/madhouse
.html. Other sources include Stacy Horn,
*Damnation Island: Poor, Sick, Mad & Crimi-
nal in 19th-Century New York* (Chapel Hill,
NC: Algonquin Books, 2018); and Matthew Goodman, *Eighty Days: Nellie Bly and
Elizabeth Bisland's History-Making Race
Around the World* (New York: Ballantine,
2013).

"The strain of playing crazy" . . . Bly, *Ten
Days in a Mad-House,* chapter 2.

"plain and unvarnished" . . . Bly, *Ten Days*

in a Mad-House, chapter 1.

two broad categories of "idiocy" and "insanity" . . . For a concise summary of the government's tracking of mental illness in America, see Herb Kutchins and Stuart A. Kirk, *Making Us Crazy* (New York: Free Press, 1997).

seven categories of mental disease . . . Allan V. Horwitz and Gerald N. Grob, "The Checkered History of American Psychiatric Epidemiology," *Milbank Quarterly* 89, no. 4 (2011): 628–57.

something called unitary psychosis . . . For more on unitary psychosis and the history of diagnosis, see Per Bergsholm, "Is Schizophrenia Disappearing? The Rise and Fall of the Diagnosis of Functional Psychoses," *BMC Psychiatry* 16 (2016): 387, https://www.ncbi.nlm.nih.gov/pmc/articles/PMC5103459.

"Compulsive epilepsy, metabolic disorders" . . . Patton State Hospital Museum, Patton, California, October 29, 2016. Thank you to curator Anthony Ortega for the enlightening tour.

Other hospital records show . . . The "other hospital" is Agnews State Hospital. The reference to "habitual consumption of peppermint candy" and "excessive tobacco use" came from Michael Svanevik and

Shirley Burgett, "Matters Historical: Santa Clara's Hospital of Horror, Agnews," *Mercury News,* October 5, 2016, https://www.mercurynews.com/2016/10/05/spdn0916matters.

were diagnosed with "insurgent hysteria" . . . Marconi Transatlantic Wireless Telegraph to the *New York Times,* "Militant Women Break Higher Law," *New York Times,* March 31, 1912, https://timesmachine.nytimes.com/timesmachine/1912/03/31/100358259.pdf.

A nineteenth-century Louisiana physician . . . Dr. Cartwright, "Diseases and Peculiarities of the Negro Race," *Africans in America,* PBS.org, https://www.pbs.org/wgbh/aia/part4/ 4h3106t.html. Thank you to Dominic Sisti and Gary Greenberg for calling my attention to these disorders.

Throw a rock into a crowd . . . For a great summary of the literature coming out of England focusing on fears about institutionalization, see Sarah Wise, *Inconvenient People: Lunacy, Liberty and the Mad-Doctors in England* (Berkeley: Counterpoint Press, 2012).

There was Lady Rosina . . . For more on Lady Rosina, see Scull, *Madness in Civilization,* 240–41.

"Never was a more criminal" . . . Rosina Bulwer Lytton, *A Blighted Life* (London: Thoemmes Press, 1994).

Elizabeth Packard continued . . . For more on Elizabeth Packard, see Linda V. Carlisle, *Elizabeth Packard: A Noble Fight* (Champaign: University of Illinois Press, 2010); and "The Case of Mrs. Packard and Legal Commitment," NIH: US National Library of Medicine, October 2, 2014, https://www.nlm.nih.gov/hmd/diseases/debates.html. For context, see Scull, *Madness in Civilization,* 240.

"Poor child," mused Judge Duffy . . . Bly, *Ten Days in a Mad-House,* chapter 4.

or mocked as "bughouse doctors" . . . Andrew Scull, *Madhouse: A Tragic Tale of Megalomania and Modern Medicine* (New Haven: Yale University Press, 2007), 14.

Psychiatrist would become . . . Scull, *Madness in Civilization,* 12.

The word *asylum* comes . . . Thank you to Arizona State classics professor Matt Simonton for explaining the Greek and Roman origins of the word *asylum.*

The first asylums built . . . Andrew Scull, "The Asylum, the Hospital, and the Clinic," *Psychiatry and Its Discontents* (Berkeley: University of California Press, 2019).

towns in Europe, the Middle East, and the Mediterranean . . . Greg Eghigan, ed., *The Routledge History of Madness and Mental Health* (New York: Routledge, 2017), 246.

there weren't many differences among . . . The rise of asylums (and their relationship to prisons and jails) is covered beautifully in David J. Rothman, *The Discovery of the Asylum: Social Order and Disorder in the New Republic* (New York: Little, Brown, 1971).

In eighteenth-century Ireland . . . Shorter, *A History of Psychiatry,* 1–2.

Europe's oldest psychiatric hospital . . . Thank you to Bethlem Museum of the Mind for providing an in-person history of their hospital and of mental health care in general. https://museumofthemind.org.uk.

a "stout iron ring" . . . Roy Porter, *Madness: A Brief History* (Oxford: Oxford University Press, 2002), 107.

American activist Dorothea Dix . . . For more on Dix, see Margaret Muckenhoupt, *Dorothea Dix: Advocate for Mental Health Care* (Oxford: Oxford University Press, 2004). For the loveliest description of her work and legacy, read Ron Powers, *No One Cares About Crazy People* (New York:

Hachette, 2017), 102–3.

thirty thousand miles across America . . .
"Dorothea Dix Begins Her Crusade,"
Mass Moments, https://www.massmo
ments.org/moment-details/dorothea-dix
-begins-her-crusade.html.

"the saddest picture of human suffering" . . .
Thomas J. Brown, *Dorothea Dix: New
England Reformer* (Boston: Harvard Uni-
versity Press, 1998), 88.

a woman tearing off her own skin . . .
Brown, *Dorothea Dix,* 89.

"sacred cause" . . . Dorothea Dix, "Memorial to the Massachusetts Legislature,
1843."

thirty-two new therapeutic asylums . . .
"Dorothea Dix Begins Her Crusade,"
Mass Moments.

"beacon for all the world" . . . Horn, *Damnation Island,* 7.

located on 147 acres . . . Horn, *Damnation
Island,* xxii.

"The mentally sick, far from being guilty
people" . . . John M. Reisman, *A History of
Clinical Psychology,* 2nd ed. (Milton Park,
UK: Taylor & Francis, 1991), 12.

Connecticut physician Eli Todd . . . The
description of his philosophy came from
Stephen Purdy, "The View from Hartford:

The History of Insanity, Shameful to Treatable," *New York Times,* September 20, 1998, https://www.nytimes.com/1998/ 09/20/nyregion/the-view-from-hartford -the-history-of-insanity-shameful-to -treatable.html.

and its "lounging, listless, madhouse air" . . . Charles Dickens, *American Notes for General Circulation* (Project Gutenberg eBook), July 18, 1998, https://www .gutenberg.org/files/675/675-h/675-h.htm. Thank you to Stacy Horn, *Damnation Island,* for making me aware of this quote.

six women were confined to a room . . . Horn, *Damnation Island,* 45.

"the onward flow of misery" . . . Horn, *Damnation Island,* 52.

give birth in a solitary cell . . . Horn, *Damnation Island,* 52.

and another woman who died . . . Horn, *Damnation Island,* 53.

"I talked and acted just as I do" . . . Bly, *Ten Days in a Mad-House,* chapter 1.

"Compare this with a criminal" . . . Bly, *Ten Days in a Mad-House,* chapter 8.

"the crib" . . . Horn, *Damnation Island,* 24.

"A human rat trap" . . . Bly, *Ten Days in a Mad-House,* chapter 16.

According to an 1874 report . . . Horn, *Damnation Island,* 16.

"more I endeavored to assure them" . . . Bly, *Ten Days in a Mad-House,* chapter 16.

"What are you doctors here for?" . . . Bly, *Ten Days in a Mad-House,* chapter 16.

The Manhattan DA convened a grand jury . . . Goodman, *Eighty Days,* 34.

"these experts cannot really tell" . . . "Nellie Brown's Story," *New York World,* October 10, 1887: 1, http://sites.dlib.nyu.edu/undercover/sites/dlib.nyu.edu.undercover/files/documents/uploads/editors/Nellie-Browns-Story.pdf.

3: The Seat of Madness

For great summaries of the early treatments of madness, see Scull, *Madness in Civilization;* Porter, *Madness: A Brief History;* Richard Noll, *American Madness: The Rise and Fall of Dementia Praecox* (Cambridge, MA: Harvard University Press, 2011); Jeffrey A. Lieberman, *Shrinks: The Untold Story of Psychiatry* (New York: Little, Brown, 2015); and of course Shorter, *A History of Psychiatry.*

unearthed skulls dated to around 5000 BC . . . Porter, *Madness: A Brief History,* 10.

Another way to rid oneself . . . Melanie

Thernstrom, *The Pain Chronicles: Cures, Myths, Mysteries, Prayers, Diaries, Brain Scans, Healing, and the Science of Suffering* (New York: FSG, 2010), 33.

"she who seizes" . . . Thernstrom, *The Pain Chronicles,* 33.

"the Lord shall smite thee" . . . Deuteronomy 28:28, the Holy Bible, King James Version (American Bible Society, 1999).

God punishes Nebuchadnezzar . . . I first encountered the story of Nebuchadnezzar in Joel Gold and Ian Gold, *Suspicious Minds: How Culture Shapes Madness* (New York: Free Press, 2014).

"those who walk in pride he is able to abase" . . . Daniel 4:37, the Holy Bible, King James Version (American Bible Society, 1999).

Those who survived suicide attempts . . . Allen Frances, *Saving Normal* (New York: William Morrow, 2013), 47.

"unambiguously a legitimate object" . . . Porter, *Madness: A Brief History,* 58.

German physician Johann Christian Reil . . . For more on Johann Christian Reil and early *psychiatrie,* see Maximilian Schochow and Florian Steger, "Johann Christian Reil (1759–1813): Pioneer of Psychiatry, City Physician, and Advocate of Public Medical Care," *American Journal of*

Psychiatry 171, no. 4 (April 2014), https://ajp.psychiatryonline.org/doi/pdfplus/10.1176/appi.ajp.2013.13081151; and Andreas Marneros, "Psychiatry's 200th Birthday," *British Journal of Psychiatry* 193, no. 1 (July 2008): 1–3, https://www.cambridge.org/core/journals/the-british-journal-of-psychiatry/article/psychiatrys-200th-birthday/6455A01CEF979FEFAB23B8467B95A823/core-reader#top.

"We will never find pure mental" . . . Quote from Marneros, "Psychiatry's 200th Birthday."

spinning chairs . . . Esther Inglis-Arkell, "The Crazy Psychiatric Treatment Developed by Charles Darwin's Grandfather," io9.gizmodo.com, July 15, 2013, https://io9.gizmodo.com/the-crazy-psychiatric-treatment-developed-by-charles-da-714873905.

"baths of surprise" . . . Andrew Scull, *Madness: A Very Short Introduction* (Oxford: Oxford University Press, 2011), 35.

Benjamin Rush, a signer of the Declaration of Independence . . . If you're interested in reading a far more flattering and nuanced portrait of Benjamin Rush, check out Stephen Fried, *Rush: Revolution, Madness, and the Visionary Doctor Who Became* a *Founding Father* (New York: Crown, 2018).

In 1874, German physician Carl Wernicke . . . Wernicke's aphasia description came from "Wernicke's (Receptive) Aphasia," National Aphasia Association, https://www.aphasia.org/aphasia-resources/wernickes-aphasia.

Frankfurt-based Dr. Alois Alzheimer . . . For more on Alois Alzheimer and his work, see Joseph Jebelli, *In Pursuit of Memory: The Fight Against Alzheimer's* (New York: Little, Brown, 2017).

though seeing a resurgence . . . "Syphilis," Sexually Transmitted Disease Surveillance 2017, CDC.gov, July 24, 2018, https://www.cdc.gov/std/stats17/syphilis.htm.

"the most destructive of all diseases" . . . John Frith, "Syphilis — Its Early History and Treatment Until Penicillin, and the Debate on Its Origins," *Journal of Military and Veterans' Health* 20, no. 4 (November 2012), https://jmvh.org/wp-content/uploads/2013/03/ Frith.pdf.

two researchers identified spiral-shaped bacteria . . . Joseph R. Berger and John E. Greenlee, "Neurosyphilis," *Neurology Medlink* (February 23, 1994), http://www.medlink.com/article/neurosyphilis.

tertiary syphilis . . . The description of

syphilis and its eventual cure came from a variety of sources, chief among them Elliot Valenstein, *Great and Desperate Cures: The Rise and Decline of Psychosurgery and Other Radical Treatments for Mental Illness* (New York: Basic Books, 1986); and Jennifer Wallis, "Looking Back: This Fascinating and Fatal Disease," *The Psychologist* 25, no. 10 (October 2012), https://thepsychologist.bps.org.uk/volume-25/edition-10/looking-back-fascinating-and-fatal-disease.

the great pox . . . Gary Greenberg, *Manufacturing Depression: The Secret History of a Modern Disease* (New York: Simon & Schuster, 2010), 55.

the infinite malady . . . "Shakespeare: The Bard at the Bedside" (editorial), *Lancet* 387 (April 23, 2016), https://www.thelancet.com/action/showPdf?pii=S0140-6736%2816%2930301-4.

the lady's disease . . . Wallis, "Looking Back."

the great imitator . . . Valenstein, *Great and Desperate Cures,* 32.

the great masquerader . . . Thank you to Dr. Heather Croy for cluing me in to this description of syphilis.

"kind of peeling" . . . Chris Frith, phone interview, August 22, 2016.

"claimed exclusive dominion" . . . Noll, *American Madness,* 17.

like stroke, multiple sclerosis, and Parkinson's . . . Mary G. Baker, "The Wall Between Neurology and Psychiatry," *British Medical Journal* 324, no. 7352 (2002): 1468–69, https://www.ncbi.nlm.nih.gov/pmc/articles/PMC1123428/.

"that could not be satisfactorily specified" . . . Noll, *American Madness,* 17.

like schizophrenia, depression, and anxiety disorders . . . Baker, "The Wall Between Neurology and Psychiatry," 1469.

German psychiatrist Emil Kraepelin . . . In addition to the many people I spoke to about Emil Kraepelin, including Andrew Scull, E. Fuller Torrey, William Carpenter, Gary Greenberg, and Ken Kendler, I credit the following sources for putting him into historical perspective: Noll, *American Madness;* and Hannah Decker, *The Making of the DSM-III: A Diagnostic Manual's Conquest of American Psychiatry* (Oxford: Oxford University Press, 2013).

This culminated in the description . . . Kraepelin did not introduce *dementia praecox* (that honor belongs to French psychiatrist Bénédict Augustin Morel), but his work clarified the term and made it widely accepted in the field.

"incurable and permanent disability" . . . Noll, *American Madness,* 66.

Swiss psychiatrist Paul Eugen Bleuler . . . For a short summary of Bleuler's contribution to psychiatry, see Paolo Fusar-Poli and Pierluigi Politi, "Paul Eugen Bleuler and the Birth of Schizophrenia (1908)," *American Journal of Psychiatry,* published online November 1, 2008, https://ajp .psychiatryonline.org/doi/10.1176/appi.ajp .2008.08050714.

psychiatrist Kurt Schneider . . . For more on Schneider's first rank symptoms, see J. Cutting, "First Rank Symptoms of Schizophrenia: Their Nature and Origin," *History of Psychiatry* 26, no. 2 (2015): 131–46, https://doi.org/10.1177/09571 54X14554369.

An American psychiatrist named Henry Cotton . . . For more on Henry Cotton, see Scull, *Madhouse.*

the growing eugenics movement . . . For more on the eugenics movement, mental illness, and sterilization, see Adam Cohen, *Imbeciles: The Supreme Court, American Eugenics, and the Sterilization of Carrie Buck* (New York: Penguin, 2017).

thirty-two states passed forced sterilization laws . . . Lisa Ko, "Unwanted Sterilization and the Eugenics Movement in the United States," *Independent Lens,* January 26, 2016, http://www.pbs.org/independent lens/blog/unwanted-sterilization-and -eugenics-programs-in-the-united-states/.

sterilizing three hundred thousand or so . . . E. Fuller Torrey and Robert H. Yolken, "Psychiatric Genocide: Nazi Attempts to Eradicate Schizophrenia," *Schizophrenia Bulletin* 36, no. 1 (January 2010): 26–32, https://www.ncbi.nlm.nih.gov/pmc/ articles/PMC2800142.

the most common diagnosis was "feeble-mindedness" . . . "Forced Sterilization," *United States Holocaust Memorial Museum,* https://www.ushmm.org/learn/students/ learning-materials-and-resources/mental ly-and-physically-handicapped-victims-of -the-nazi-era/forced-sterilization.

especially in 1955, when over a half million people . . . Andrew Scull, *Decarceration: Community Treatment and the Deviant — A Radical View* (Englewood Cliffs, NJ: Prentice Hall, 1977), 80.

Psychoanalysis invaded the US . . . For more on psychoanalysis in the United States, see Janet Malcolm, *Psychoanalysis: The Impossible Profession* (New York: Vintage Books, 1980); Jonathan Engel, *American Therapy: The Rise of Psychother-*

apy in the United States (New York: Gotham Books, 2008); and T. M. Luhrmann, *Of Two Minds: An Anthropologist Looks at American Psychiatry* (New York: Vintage, 2001).

"nothing arbitrary or haphazard" . . . Malcolm, *Psychoanalysis,* 19.

German judge Daniel Paul Schreber . . . Information on Schreber was gathered from Thomas Dalzell, *Freud's Schreber: Between Psychiatry and Psychoanalysis* (London: Karnac Books, 2011).

"a power, a secular power" . . . Allen Frances, phone interview, January 4, 2016.

"family relations, cultural traditions, work patterns" . . . Bonnie Evans and Edgar Jones, "Organ Extracts and the Development of Psychiatry: Hormonal Treatments at the Maudsley Hospital, 1923–1938," *Journal of Behavioral Science* 48, no. 3 (2012): 251–76.

The people who needed help the most . . . Freud, it should be noted, did not believe that psychoanalysis worked on people with schizophrenia. "Freud thought that because of the nature of the libidinal withdrawal in schizophrenia and paranoia, the patient could not form a transference and thus could not be treated." William N. Goldstein, "Toward an Integrated Theory of Schizophrenia," *Schizophrenia Bulletin* 4, no. 3 (January 1978): 426–35, https://academic.oup.com/schizophreniabulletin/article-abstract/4/3/426/1874808.

Freud's nephew Edward Bernays . . . For more on Freud's nephew Edward Bernays and the use of Freud's theories by corporations and government, see Adam Curtis, *The Century of the Self* (documentary), British Broadcasting Corporation, 2006.

"interchange of words" . . . Sigmund Freud, "First Lecture: Introduction," in *A General Guide to Psychoanalysis* (New York: Boni and Liveright, 1920), https://www.bartleby.com/283/.

"the most complex of the talking treatments" . . . "Psychoanalysis and Psychotherapy," British Psychoanalytic Council, https://www.bpc.org.uk/psychoanalysis-and-psychotherapy.

Viennese psychoanalyst Bruno Bettelheim . . . Bruno Bettelheim, *The Empty Fortress: Infantile Autism and the Birth of the Self* (New York: Free Press, 1972).

"psychoanalyst of vast impact" . . . Daniel Goleman, "Bruno Bettelheim Dies at 86; Psychoanalyst of Vast Impact," *New York Times,* March 14, 1990, https://www.nytimes.com/1990/03/14/obituaries/bruno-bettelheim-dies-at-86-psychoanalyst-of

-vast-impact.html.

allegations emerged that Bettelheim . . . Joan Beck, "Setting the Record Straight About a 'Fallen Guru,' " *Chicago Tribune,* April 3, 1997, https://www.chicagotribune .com/news/ct-xpm-1997-04-03-970403 0057-story.html.

"extreme diagnostic nihilism" . . . David Healy, *The Antidepressant Era* (Cambridge, MA: Harvard University Press, 2014), 41.

"true mental health was an illusion" . . . Luhrmann, *Of Two Minds,* 218.

a now infamous 1962 Midtown Manhattan study . . . Leo Srole, Thomas S. Langner, Stanley T. Michael, et al., *Mental Health in the Metropolis: The Midtown Manhattan Study* (New York: McGraw-Hill), 1962.

4: On Being Sane in Insane Places

The meeting with Dr. Deborah Levy and Dr. Joseph Coyle took place on March 20, 2013. Thank you to Brookline Booksmith for inviting me to Boston and making this meeting possible.

"For ten days I had been one of them" . . . Nellie Bly, "Among the Mad," *Godey's Lady's Book,* January 1889, https://www .accessible-archives.com/2014/05/nellie

-bly-among-the-mad.

"The facts of the matter are" . . . David Rosenhan, "On Being Sane in Insane Places," *Science* 179, no. 4070 (January 19, 1973): 257.

"like a sword plunged" . . . Robert Spitzer, "Rosenhan Revisited: The Scientific Credibility of Lauren Slater's Pseudopatient Diagnosis Study," *Journal of Nervous and Mental Disease* 193, no. 11 (November 2005).

"If sanity and insanity exist" . . . Rosenhan, "On Being Sane in Insane Places," 250.

"essentially eviscerated any vestige" . . . Jeffrey Lieberman, phone interview, February 25, 2016.

"Psychiatrists looked like unreliable" . . . Frances, *Saving Normal,* 62.

nearly 80 percent of all intro-to-psychology textbooks . . . Jared M. Bartels and Daniel Peters, "Coverage of Rosenhan's 'On Being Sane in Insane Places' in Abnormal Psychology Textbooks," *Society for the Teaching of Psychology* 44, no. 2 (2017): 169–73.

Rosenhan study takes up nearly a whole page . . . Tom Burns, *Psychiatry: A Very Short Introduction* (Oxford: Oxford University Press, 2006), 114.

"a bunch of harum-scarum sensationalists" . . . Ed Minter, "Still Inexact Science," *Albuquerque Journal,* January 29, 1973.

Eight people — Rosenhan himself and seven others . . . All details about the study here are from Rosenhan, "On Being Sane in Insane Places."

"Each was told" . . . Rosenhan, "On Being Sane in Insane Places," 252.

30 percent of fellow patients . . . More specifically, 35 of a total 118 patients encountered voiced suspicions. Rosenhan, "On Being Sane in Insane Places."

"You're not crazy" . . . Rosenhan, "On Being Sane in Insane Places," 252.

"patient engages in writing behavior" . . . Rosenhan, "On Being Sane in Insane Places," 253.

"Having once been labeled schizophrenic" . . . Rosenhan, "On Being Sane in Insane Places," 253.

"How many people, one wonders" . . . Rosenhan, "On Being Sane in Insane Places," 257.

Science's most famous papers include . . . For a brief history of *Science,* see "About Science & AAAS," https://www.science mag.org/about/about-science-aaas?r3f _986=https://www.google.com.

In 1971, a large-scale US/UK study

showed . . . Robert E. Kendell, John E. Cooper, Barry J. Copeland, et al., "Diagnostic Criteria of American and British Psychiatrists," *Archives of General Psychiatry* 25, no. 2 (August 1971): 123–30.

concluding in his 1962 paper . . . Aaron T. Beck, "Reliability of Psychiatric Diagnoses: A Critique of Systematic Studies," *American Journal of Psychiatry* 119 (1962): 210–16.

state hospitals had released half . . . E. Fuller Torrey, "Ronald Reagan's Shameful Legacy: Violence, the Homeless, Mental Illness," *Salon,* September 29, 2013, https://www.salon.com/2013/09/29/ronald _reagans_shameful_legacy_violence_the _homeless_mental_illness/.

Various Reddit pages . . . One example has forty-three hundred comments as of April 1, 2019: https://www.reddit.com/r/ todayilearned/comments/6qzaz1/til_about _the_rosenhan_experiment_in_which_a.

one college student at Jacksonville State Hospital . . . John Power, "Find Pseudo-Patient at State Hospital," *Jacksonville Daily Journal,* May 9, 1973.

He testified in a Navy hearing . . . Messrs. Vernon Long, John Wherry, and Walter Champion, Navy Board of Investigation, Cong. 1–50 (1973) (testimony of David

Rosenhan, PhD), David L. Rosenhan Papers (SC1116), Department of Special Collections and University Archives, Stanford University Libraries, Stanford, California.

"flipping coins" . . . Bruce J. Ennis and Thomas R. Litwack, "Psychiatry and the Presumption of Expertise: Flipping Coins in the Courtroom," *California Law Review* 62, no. 693 (1973).

"It is not known why powerful impressions" . . . Rosenhan, "On Being Sane in Insane Places," 254.

"At times, depersonalization reached" . . . Rosenhan, "On Being Sane in Insane Places," 256.

"Rather than acknowledge" . . . Rosenhan, "On Being Sane in Insane Places," 257.

5: A Riddle Wrapped in a Mystery Inside an Enigma

The bulk of this chapter came from my visit with Professor Lee Ross at his office at Stanford on November 3, 2015, where he granted my request for a personal interview.

he did reassure the superintendent . . . David Rosenhan, letter to Dr. Kurt Anstreicher, March 15, 1973.

"Through the publicity" . . . Paul R. Fleischman, letter to the editor, *Science,* April 27, 1973: 356, http://science.sciencemag .org/content/180/4084/356.

"It can only be productive of" . . . Otto F. Thaler, letter to the editor, *Science,* April 27, 1973: 358.

Lauren Slater claimed that . . . Lauren Slater, "On Being Sane in Insane Places," in *Opening Skinner's Box: Great Psychological Experiments of the Twentieth Century* (New York: W. W. Norton, 2004).

A BBC radio report . . . Claudia Hammond, "The Pseudo-Patient Study," *Mind Changers,* BBC Radio 4, July 2009, https:// www.bbc.co.uk/programmes/b00lny48.

his close friend and colleague Lee Ross . . . For more on Lee Ross's contributions to psychology, see his seminal work (recently re-released with a foreword by Malcolm Gladwell): Lee Ross and Richard Nisbett, *The Person and the Situation: Perspectives of Social Psychology* (New York: McGraw-Hill, 1991).

as widespread as left-handedness . . . About 12 percent of the population is left-handed, and studies have shown that the median prevalence of auditory hallucinations in the general public is 13.2 percent. Louis C. Johns, Kristiina Kompus, Melissa

Connell, et al., "Auditory Verbal Hallucinations in Persons Without a Need for Care," *Schizophrenia Bulletin* 40, no. 4 (2014): 255–64, https://academic.oup.com/schizophreniabulletin/article/40/Suppl_4/S255/1873600.

you're joining an esteemed group . . . Joe Pierre, "Is It Normal to 'Hear Voices'?" *Psychology Today,* August 31, 2015, https://www.psychologytoday.com/us/blog/psych-unseen/201508/is-it-normal-hear-voices.

the much-publicized Stanford Prison Experiment . . . Craig Haney, Curtis Banks, and Philip Zimbardo, "Interpersonal Dynamics in a Simulated Prison," *International Journal of Criminology and Penology* 1 (1973): 69–97, http://pdf.prisonexp.org/ijcp1973.pdf.

"pseudoscience presented as science" . . . Robert Spitzer, "On Pseudoscience in Science, Logic in Remission, and Psychiatric Diagnosis: A Critique of Rosenhan's 'On Being Sane in Insane Places,' " *Journal of Abnormal Psychology* 84, no. 5 (1975): 442–52.

"unfounded" . . . Bernard Weiner, " 'On Being Sane in Insane Places': A Process (Attributional) Analysis and Critique," *Journal of Abnormal Psychology* 84, no. 5

(1975): 433–41.

"entirely unwarranted" . . . George Weideman, "Psychiatric Disease: Fiction or Reality?" *Bulletin of the Menninger Clinic* 37, no. 5 (1973): 519–22.

The excerpt from chapter 1 came from David Rosenhan's unpublished book *Odyssey into Lunacy,* from his personal files.

Pseudopatient list compiled from pseudopatient notes and the unpublished book located in David Rosenhan's personal files.

Part Two

Felix Unger: I think I'm crazy . . . *The Odd Couple,* directed by Gene Sacks, Paramount Pictures, 1968.

6: The Essence of David

Florence is trim and attractive . . . I learned about David Rosenhan and Florence Keller's history over many lovely interviews, but this meeting happened the week of June 14, 2014.

"It all started out as a dare" . . . David Gunter, "Study of Mental Institutions Began as a Dare," *Philadelphia Daily News,* January 19, 1973.

January 1969, Swarthmore, Pennsylvania . . . Details about Swarthmore in the late 1960s were compiled via many sources — most notably David's unpublished book *Odyssey into Lunacy*. I also visited Swarthmore College and accessed their archives, which had a few documents relating to David's hiring and eventual move to Stanford. In addition, the 1969 and 1970 Halcyon yearbook and student newspaper the *Phoenix* provided colorful context. I also turned to other secondary sources to put together a wider snapshot of this time in American history: Clara Bingham, *Witness to the Revolution: Radicals, Resisters, Vets, Hippies, and the Year America Lost Its Mind and Found Its Soul* (New York: Random House, 2017); *The Sixties* (mini-series), produced by Tom Hanks and Playtone, CNN, 2014; Rob Kirkpatrick, *1969: The Year Everything Changed* (New York: Skyhorse Publishing, 2011); Andreas Hillen, *1973 Nervous Breakdown: Watergate, Warhol, and the Birth of Post-Sixties America* (New York: Bloomsbury, 2006); Brendan Koerner, *The Skies Belong to Us: Love and Terror in the Golden Age of Hijacking* (New York: Crown, 2013); Todd Gitlin, *The Sixties:*

Years of Hope, Days of Rage (New York: Bantam, 1988); and Jules Witcover, *The Year the Dream Died: Revisiting 1968 in America* (New York: Grand Central, 1997).

more than eighty-four incidences of bombings . . . Kirkpatrick, *1969,* 14.

Richard Nixon's inauguration . . . To see more about Nixon's inauguration, see "1968," *The Sixties,* CNN.

casualties hit their peak in 1968 . . . There were nearly 16,889 deaths in 1968. "Vietnam War U.S. Military Fatal Casualty Statistics: Electronic Records Report," National Archives, https://www.archives .gov/research/military/vietnam-war/ casualty-statistics#date.

"It's easy to forget" . . . Mark Vonnegut, *The Eden Express: A Memoir of Insanity* (New York: Seven Stories Press, 2002), 15.

"Lose your mind and come to your senses" . . . According to several books on Gestalt therapy, "Lose your mind and come to your senses" (or variations on this theme) was one of Fritz Perls's favorite sayings.

Two million Americans . . . Bingham, *Witness to the Revolution,* xxviii.

as Joan Didion wrote . . . Joan Didion, *The*

White Album (New York: Farrar Straus and Giroux, 2009 edition), 121.

One of the country's most popular bumper stickers . . . Bingham, *Witness to the Revolution,* 432.

Ken Kesey's trippy novel . . . The following sources were helpful in putting together a short sketch of Ken Kesey and *One Flew Over the Cuckoo's Nest:* Robert Faggen, introduction to *One Flew Over the Cuckoo's Nest,* 4th ed. (New York: Penguin Books, 2002), ix–xxv; James Wolcott, "Still *Cuckoo* After All These Years," *Vanity Fair,* November 18, 2011, http://www.vanityfair .com/news/2011/12/wolcott-201112; Nathaniel Rich, "Ken Kesey's Wars: 'One Flew Over the Cuckoo's Nest' at 50," *Daily Beast,* July 26, 2012, https://www .thedailybeast.com/ken-keseys-wars-one -flew-over-the-cuckoos-nest-at-50.

"gave life to a basic distrust" . . . Jon Swaine, "How 'One Flew Over the Cuckoo's Nest' Changed Psychiatry," *The Telegraph,* February 1, 2011, https://www .telegraph.co.uk/news/worldnews/north america/usa/8296954/How-One-Flew -Over-the-Cuckoos-Nest-changed-psychi atry.html.

"If it gets me outta those damn pea

fields" . . . Kesey, *One Flew Over the Cuckoo's Nest,* 13.

"Hell, I been surprised" . . . Kesey, *One Flew Over the Cuckoo's Nest,* 58.

"I discovered at an early age" . . . Kesey, *One Flew Over the Cuckoo's Nest,* 265.

Cold War paranoia touched everyone . . . For more on the abuses of Soviet Union psychiatry, see Richard Bentall, *Madness Explained: Psychosis and Human Nature* (New York: Penguin Books, 2004); and Robert van Voren, "Political Abuse of Psychiatry — An Historical Overview," *Schizophrenia Bulletin* 36, no. 1 (January 2010): 33–35, https://doi.org/10.1093/schbul/sbp119.

outspoken general named Pyotr Grigorenko . . . I first encountered the story of Pyotr Grigorenko in David Rosenhan's own writing: David Rosenhan, "Psychology, Abnormality and Law," Master Lecture in Psychology and Law, presented at the Meeting of the American Psychological Association, Washington, DC, August 1982 (found in David Rosenhan's personal files). For more on Grigorenko, see W. Reich, "The Case of General Grigorenko: A Psychiatric Reexamination of a Soviet Dissident," *Psychiatry* 43, no. 4 (1980): 303–23; and James Barron, "Petro

Grigorenko Dies in Exile in US," *New York Times,* February 23, 1987, https://www.nytimes.com/1987/02/23/obituaries/petro-grigorenko-dies-in-exile-in-us.html.

He spent five years . . . "Pyotr G. Grigorenko, Exiled Soviet General, Dies in N.Y." *Los Angeles Times,* February 25, 1987, https://www.latimes.com/archives/la-xpm-1987-02-25-mn-5733-story.html.

"a dangerous lunatic" . . . "1,189 Psychiatrists Say Goldwater Is Psychologically Unfit to Be President!," which ran in *Fact* magazine in 1964.

"Psychiatrists are medical doctors" . . . American Psychiatric Association, "APA Calls for End to 'Armchair' Psychiatry,' " Psychiatry.org, January 9, 2018, https://www.psychiatry.org/newsroom/news-releases/apa-calls-for-end-to-armchair-psychiatry.

R. D. Laing, a Scottish psychiatrist . . . To understand R. D. Laing in full, you need to read his work, but I also highly recommend reading his son's biography of him: Adrian Laing, *R. D. Laing: A Life* (New York: Pantheon Books, 1997).

"They will see that" . . . R. D. Laing, *The Politics of Experience* (New York: Random House, 1967), 107.

"Madness need not be all breakdown" . . .

Laing, *The Politics of Experience,* 133.

"Schizophrenics were the true poets" . . . Erica Jong, *Fear of Flying* (New York: Penguin Books, 1973), 82.

Thomas Szasz called mental illnesses a "myth" . . . Thomas Szasz, preface to *The Myth of Mental Illness* (1961; 2nd reissue, Harper Perennial, 2003).

"If you talk to God" . . . Thomas Szasz, *The Second Sin* (Garden City, NY: Anchor Press, 1973), 101.

whom he called "parasites" . . . Thomas Szasz, *Cruel Compassion: Psychiatric Control of Society's Unwanted* (New York: Wiley, 1994), 142.

"The Crisis of 1969" . . . Material was pieced together from articles published in the student newspaper the *Phoenix,* specifically Russ Benghiat, Doug Blair, and Bob Goodman, "Crisis of '69: Semester of Misunderstanding and Frustration," *Swarthmore College Phoenix,* January 29, 1969: 4–6. A more recent examination can be found in Elizabeth Weber, "The Crisis of 1969," *Swarthmore College Phoenix,* March 7, 1996, http://www.sccs.swarthmore.edu/users/98/elizw/Swat.history/69.crisis.html; and Kirkpatrick, *1969,* 10–11.

Vice President Spiro Agnew . . . This came

559

from my interview with Swarthmore psychology professor Barry Schwartz and has been repeated in various articles. A recent Swarthmore student newspaper article, however, casts some doubt that Spiro Agnew coined the phrase. Miles Skorpen, "Where Does the 'Kremlin on the Crum' Come From?" *The Phoenix,* March 6, 2007, https://swarthmorephoenix.com/2007/03/06/ask-the-gazette-where-does -the-kremlin-on-the-crum-come-from/.

7: "Go Slowly, and Perhaps Not at All"

This chapter was pieced together with help from Rosenhan's unpublished book and interviews with Jack Rosenhan, Florence Keller, and former students.

Rosenhan was a scrawny kid . . . Jack Rosenhan, in-person interview, October 21, 2015.

dream analysis . . . Edith Sheppard and David Rosenhan, "Thematic Analysis of Dreams," *Perceptual and Motor Skills* 21 (1965): 375–84.

hypnosis . . . David Rosenhan, "On the Social Psychology of Hypnosis Research," in Jesse E. Gordon, ed., *Handbook of Clinical and Experimental Hypnosis* (New York: Macmillan, 1967), 481–510.

Freedom Riders . . . David Rosenhan, "Determinants of Altruism: Observations for a Theory of Altruistic Development," paper presented at an annual meeting of the American Psychological Association, September 1969, https://files.eric.ed.gov/fulltext/ED035035.pdf.

He replicated Stanley Milgram's 1963 study . . . David Rosenhan, "Obedience and Rebellion: Observations on the Milgram Three Party Paradigm," Draft, November 27, 1968, David L. Rosenhan Papers.

Milgram had created a fake shock box . . . For more on Milgram (and subsequent questions about his research), see Gina Perry, *Behind the Shock Machine: The Untold Story of the Notorious Milgram Psychology Experiments* (New York: New Press, 2013).

"A number of us here" . . . David Rosenhan, letter to Stanley Milgram, July 9, 1963, Milgram Papers, Series III, Box 55, Folder 12.

"young children's unprompted concern" . . . Rosenhan wrote many articles about altruism and children, among them David Rosenhan and Glenn M. White, "Observation and Rehearsal as Determinants of Prosocial Behavior," *Journal of Personality and Social Psychology* 5, no. 4 (1967): 424–31; David Rosenhan, "The Kindnesses of Children," *Young Children* 25, no. 1 (October 1969): 30–44; and David Rosenhan, "Double Alternation in Children's Binary Choice," *Psychonomic Science* 4 (1966): 431–32.

Rosenhan set up his lab . . . The descriptions of Rosenhan's lab came from his unpublished book; a log of all the equipment bought for his lab; descriptions from interviews with two of his lab assistants, Bea Patterson and Nancy Horn; and his academic papers.

then documented how the child's altruistic behavior . . . Rosenhan and White, "Observation and Rehearsal as Determinants of Prosocial Behavior."

published another, more interesting paper about the role of confidence . . . Alice M. Isen, Nancy Horn, and David L. Rosenhan, "Effects of Success and Failure on Childhood Generosity," *Journal of Personality and Social Psychology* 27, no. 2 (1973): 239–47.

"Abnormal psychology is a painfully complicated" . . . David Rosenhan, September 12, 1972, David L. Rosenhan Papers.

"rivet a group of two to three hundred students" . . . Pauline Lord, letter to David Rosenhan. April 5, 1973, David L. Rosenhan Papers.

"The question is . . . What is abnormality?" . . . David Rosenhan, abnormal psychology class lectures (cassette), Stanford University, undated.

"that the course had had two shortcomings" . . . David Rosenhan, *Odyssey into Lunacy,* chapter 1, 2.

an undergraduate course at Yeshiva University . . . Description of Yeshiva University and the minority groups class came from Rosenhan's unpublished book.

Kremens, who had worked . . . This and other details about Kremens were learned in an in-person interview with his son and Mrs. Kremens on April 12, 2017.

a Haverford Hospital nurse named Linda Rafferty . . . Susan Q. Stranahan, "Ex-Haverford Nurse Sues to Regain Job," *Philadelphia Inquirer,* December 30, 1972.

"homosexual abuse by other patients" . . . *Commonwealth of Pennsylvania ex rel. Linda Rafferty et al. v. Philadelphia Psychiatric Center et al.,* 356 F. Supp. 500, United States District Court, March 27, 1973.

"the first drug that worked" . . . Shorter, *A History of Psychiatry,* 246.

"widely cited as rivaling penicillin" . . . David Healy, *Pharmageddon* (Berkeley: University of California Press, 2012), 88.

"Thousands of patients who had been assaultive" . . . Susan Sheehan, *Is There No Place on Earth for Me?* (New York: Houghton Mifflin Harcourt, 1982), 10.

to the tune of $116.5 million . . . Scull, *Decarceration,* 80.

depression was still viewed by many . . . Michael Alan Taylor, *Hippocrates Cried: The Decline of American Psychiatry* (Oxford: Oxford University Press, 2013), 19.

We developed schizophrenia . . . Healy, *The Antidepressant Era,* 162.

"Miss Ratched shall line us" . . . Kesey, *One Flew Over the Cuckoo's Nest,* 262.

"We were all keyed up" . . . Harvey Shipley Miller, phone interview, January 26, 2016.

"They will probably write a paper about it!" . . . Rosenhan, *Odyssey into Lunacy,* handwritten notes, private files.

among them medical anthropologist William Caudill . . . William Caudill, Frederick C. Redlich, Helen R. Gilmore, and Eugene B. Brody, "Social Structure and Interaction Processes on a Psychiatric Ward," *American Journal of Orthopsychiatry* 22, no. 2 (1952): 314–34, https://onlinelibrary.wiley.com/doi/pdf/10.1111/j.1939-0025.1952.tb01959.x.

"I believe he lost his objectivity" . . . Martin Bulmer, "Are Pseudo-Patient Studies Justified?" *Journal of Medical Ethics* 8 (1982): 68.

"not alter our life histories" . . . Rosenhan, *Odyssey into Lunacy,* chapter 2, 16.

During World War II, three thousand conscientious objectors . . . Joseph Shapiro, "WWII Pacifists Exposed Mental Ward Horrors," *NPR,* December 30, 2009, https://www.npr.org/templates/story/story.php?storyId=122017757.

featured in Albert Maisel's "Bedlam 1946" . . . Albert Maisel, "Bedlam 1946," *Life,* May 6, 1946, 102–18.

Harold Orlansky compared American asylums . . . Harold Orlansky, "An American Death Camp," *Politics* (1948): 162–68, http://www.unz.com/print/Politics-1948q2-00162.

Frederick Wiseman's damning documentary . . . *Titicut Follies,* directed by Frederick Wiseman, American Direct Cinema, 1967.

Goffman described the hospital as a "total institution" . . . Erving Goffman, *Asylums* (New York: Doubleday, 1961).

a condition that psychiatrist Russell Barton . . . Russell Barton, *Institutional Neurosis* (Ann Arbor: University of Michigan

Press, 1959).

"authoritarian" . . . . These three descriptions came from notes provided to me by Swarthmore student Hank O'Karma, who attended a different seminar on abnormal psychology the previous semester. The original source is J. D. Holzberg, "The Practice and Problems of Clinical Psychology in a State Psychiatric Hospital," *Journal of Consulting Psychology* 16, no. 2 (1952).

"degrading" . . . T. R. Sarbin, "On the Futility of the Proposition that Some People Be Labeled 'Mentally Ill,' " *Journal of Consulting Psychology* 31, no. 5 (1967): 447–53.

"illness-maintaining" . . . Alfred H. Stanton and Morris S. Schwartz, *The Mental Hospital: A Study of Institutional Participation in Psychiatric Illness and Treatment* (New York: Basic Books, 1954). A fun aside: Morris Schwartz is better known as the subject of Mitch Albom's *Tuesdays with Morrie: An Old Man, a Young Man, and Life's Greatest Lesson* (New York: Doubleday, 1997).

"Wasn't it dangerous?" . . . David Rosenhan, *Odyssey into Lunacy*, chapter 1, 5.

"Perhaps hospitals cure" . . . David Rosenhan, "Brief Description," private files.

"Go slowly" . . . Rosenhan, *Odyssey into Lunacy,* handwritten notes, private files.

Dr. Orne would later make waves . . . Alessandra Stanley, "Poet Told All; Therapist Provides the Record," *New York Times,* July 15, 1991, https://www.nytimes.com/1991/07/15/books/poet-told-all-therapist-provides-the-record.html.

8: "I Might Not Be Unmasked"

This chapter was compiled with help from David's unpublished book, his diary entries, and letters and correspondences exchanged around that time.

Rosenhan didn't do anything . . . Jack Rosenhan, in-person interview, October 21, 2015.

"Thinking and discussing are not" . . . Rosenhan, *Odyssey into Lunacy,* chapter 3, 1.

They had met on the first day . . . I learned about the Rosenhans' courtship thanks to various interviews with Jack Rosenhan and with Mollie's oldest friend, Abbie Kurinsky (January 14, 2014).

"Remember how I touched your arm" . . . David Rosenhan, letter to Mollie, undated.

The phone logs recorded a man . . . Haverford State Hospital medical records, February 5, 1969, David Rosenhan private papers.

He put on an old raggedy . . . Rosenhan, *Odyssey into Lunacy,* chapter 3, 5a.

Two court-martialed soldiers . . . Wallace Turner, "Sanity Inquiry Slated in Setback for Defense at Trial for Mutiny," *New York Times,* February 6, 1969, https://times machine.nytimes.com/timesmachine/ 1969/02/06/88983251.html?pageNum ber=16.

younger brother struggled with manic depression . . . Jack Rosenhan, in-person interview, October 21, 2015.

he grew even more conservative . . . Jack Rosenhan, in-person interview, October 21, 2015.

during manic phases when off his medications . . . Jack Rosenhan, in-person interview, October 21, 2015.

"My dad was constantly on the phone" . . . Jack Rosenhan, in-person interview, October 21, 2015.

Jack believed that these experiences . . . Jack Rosenhan, in-person interview, February 20, 2017.

"a fear that I might *not* be unmasked" . . .

Rosenhan, *Odyssey into Lunacy*, chapter 3, 2.

"Do I need shirts, ties, and underwear" . . . Rosenhan, *Odyssey into Lunacy*, chapter 3, 2.

A semicircular gray stone wall . . . My description of Haverford State was compiled with help from H. Michael Zal, *Dancing with Medusa: A Life in Psychiatry: A Memoir* (Bloomington, IN: Author House, 2010); and "Governor Hails New Hospital," *Delaware County Daily Times,* September 13, 1962: 1.

the Haverford Hilton . . . Zal, *Dancing with Medusa,* 12.

"the Queen Ship" . . . Mack Reed, " 'Queen Ship' of Hospitals Foundering," *Philadelphia Inquirer,* October 1, 1987, http://articles.philly.com/1987-10-01/news/26217259.

"showpiece of radical design" . . . Reed, " 'Queen Ship' of Hospitals Foundering."

British psychiatrist Humphry Osmond . . . Thanks to the following sources for their information and insight into Humphry Osmond (who is far more fascinating than I had space to describe): R. Sommer, "In Memoriam: Humphry Osmond," *Journal of Environmental Psychology* 24 (2004): 257–58; Erika Dyck, *Psychedelic Psychia-*

try (Baltimore: Johns Hopkins University Press, 2008); Tom Shroder, *Acid Test* (New York: Blue Rider, 2014); Jay Stevens, *Storming Heaven: LSD and the American Dream* (New York: Atlantic Monthly Press, 1987); Janice Hopkins Tanne, "Humphry Osmond," *British Medical Journal* 328, no. 7441 (March 2004): 713; and Michael Pollan, *How to Change Your Mind: What the New Science of Psychedelics Teaches Us About Consciousness, Dying, Addiction, Depression, and Transcendence* (New York: Penguin Press, 2018).

a "guru of the 1960s" . . . Sommer, "In Memoriam," 257.

"They're ugly monuments" . . . Sidney Katz, "Osmond's New Deal for the Insane," *Maclean's,* August 31, 1957, http:// archive.macleans.ca/article/1957/8/31/dr -osmonds-new-deal-for-the-insane.

he made the wards circular . . . Humphry Osmond, "Function as the Basis of Psychiatric Ward Design," *Mental Hospitals,* April 1957, https://ps.psychiatryonline.org/doi/ 10.1176/ps.8.4.23.

"enter the illness and see" . . . Humphry Osmond, "On Being Mad," *Saskatchewan Psychiatric Services Journal* 1, no. 1 (1952), http://www.psychedelic-library .org/ON%20BEING%20MAD.pdf.

"It would be heartless to house" . . . Osmond, "Function as the Basis of Psychiatric Ward Design."

The patterned tiles . . . P. G. Stafford and B. H. Golightly, *LSD: The Problem-Solving Psychedelic* (New York: Award Books, 1967), https://www.scribd.com/doc/12692270/LSD-The-Problem-Solving-Psychedelic.

"illusion-producing machines" . . . Stafford and Golightly, *LSD,* 208.

double-Y-shaped structure . . . Zal, *Dancing with Medusa,* 29.

"used here but not loved" . . . Rosenhan, *Odyssey into Lunacy,* chapter 3, 3.

"Not a picture nor an object" . . . Rosenhan, *Odyssey into Lunacy,* chapter 3, 3.

Case Number: #5213 . . . Haverford State Hospital medical records.

What if I had really been a patient? . . . Rosenhan, *Odyssey into Lunacy,* chapter 3, 3.

9: Committed

This chapter was compiled with the help of David Rosenhan's Haverford State Hospital medical records, his unpublished book, and interviews with Dr. Bartlett's daughter Mary (January 30, 2017) and former assistant Carole Adrienne Murphy (March 13, 2017).

hardly ever without a cigarette . . . Mary Bartlett, phone interview, January 30, 2017.

"I've been hearing voices" . . . David Rosenhan, *Odyssey into Lunacy,* chapter 3, 4–11.

"He has tended to get lost" . . . F. Lewis Bartlett, Haverford State Hospital medical records.

"This man who is unusually intelligent" . . . F. Lewis Bartlett, Haverford State Hospital medical records.

Impression . . . Excerpt from Haverford State Hospital medical records.

Should they call Jack Kremens? . . . adapted from Rosenhan, *Odyssey into Lunacy,* chapter 3.

"You both are crazy" . . . David Rosenhan, "Odyssey into Lunacy — notes on nether people," handwritten and undated, private files.

"really for the patient's own good" . . . Rosenhan, *Odyssey into Lunacy,* chapter 3, 13.

"Like hell it didn't matter!" . . . Rosenhan, *Odyssey into Lunacy,* chapter 3, 13.

"we do not administer any type" . . . Rosenhan, *Odyssey into Lunacy,* chapter 3, 13.

10: Nine Days Inside a Madhouse

I re-created David's nine-day hospitalization by pulling from his unpublished book *Odyssey into Lunacy,* his diary entries written at the time of his hospitalization, medical records, and various notes and records taken at the time. For context and description, I also added details from *Dancing with Medusa* by H. Michael Zal. All direct quotes are pulled from David's writing.

All nurses' notes are from Haverford State Hospital medical records.

because it was "illegal" . . . Rosenhan, *Odyssey into Lunacy,* chapter 3, 14.

"Opening the locked door" . . . Zal, *Dancing with Medusa,* 44.

"Son of a bitch!" . . . handwritten diary notes, undated page, Rosenhan private files.

"What the hell have I gotten myself into?" . . . Zal, *Dancing with Medusa,* 45.

Where to wash up or to shower? . . . Rosenhan, *Odyssey into Lunacy,* chapter 7, 3.

A blaring fire alarm . . . Rosenhan, *Odyssey into Lunacy,* chapter 7, 6.

"C'MON, YOU MOTHERFUCKERS, LET'S GO" . . . Rosenhan, handwritten diary entry, February 8, 1969, private files.

sniffing glue . . . Rosenhan, handwritten diary entry, February 7, 1969.

"He knew I had been watching" . . . Rosenhan, *Odyssey into Lunacy,* chapter 7, 7.

"I looked in the mirror" . . . Rosenhan, *Odyssey into Lunacy,* chapter 7, 9.

"Hey, one butter only" . . . Rosenhan, *Odyssey into Lunacy,* chapter 9, 10.

"Tom Szasz is wrong" . . . Rosenhan, diary entries, February 1969.

"Not everyone reads them" . . . Rosenhan, diary entry, February 9, 1969.

"so drugged was I from heat" . . . Rosenhan, diary entry, "Keeping their distance," undated.

"The walls here are plaster" . . . Rosenhan, diary entry, "Keeping their distance," undated.

They discussed Rosenhan's financial difficulties . . . Robert Browning, Haverford State Hospital medical records.

In 1946, Polish psychologist Solomon Asch studied . . . Solomon Asch, "Forming Impressions of Personality," *Journal of Abnormal and Social Psychology* 41, no. 3 (1946): 258–90.

two psychologists played a recorded conversation . . . E. J. Langer and R. P. Abelson, "A Patient by Another Name: Clinical Group Difference in Labeling Bias," *Jour-*

nal of Consulting and Clinical Psychology 42 (1974): 4–9.

typical outcome of "the medical gaze" . . . Michel Foucault, *The Birth of the Clinic: An Archaeology of Medical Perception* (New York: Pantheon, 1973).

"residual type," defined as a person who has exhibited signs . . . American Psychiatric Association, "Glossary of Terms," in *Diagnostic and Statistical Manual of Mental Disorders,* 2nd ed. (Washington, DC: American Psychiatric Association, 1968), 34–35.

"Have my clothes come up yet?" . . . Rosenhan, diary entries, "4pm," February 7, 1969.

"whiling it away" . . . Rosenhan, diary entries, undated.

"almost as if the disorder" . . . Rosenhan, "On Being Sane in Insane Places," 254.

"No, she was not being seductive" . . . Rosenhan, *Odyssey into Lunacy,* chapter 7, 3.

He flipped through articles . . . *New York Times,* January 31, 1969, https://timesmachine.nytimes.com/timesmachine/1969/01/31/issue.html.

"Would I have to be secretive?" . . . Rosenhan, diary entries, February 7, 1969.

"I'm Bob Harris" . . . The Bob Harris

interaction came from Rosenhan, *Odyssey into Lunacy,* chapter 7, 12–16.

"Even Harris' differentiated friendliness" . . . Rosenhan, diary entry, February 8, 1969.

wintry Sunday . . . weather found thanks to https://www.wunderground.com/history/ weekly/KPHL/date/1969-2-9?req_city= &req_state=&req_statename=&reqdb .zip=& reqdb.magic=&reqdb.wmo=.

"The pacing, the sitting" . . . Rosenhan, diary entry, February 9, 1969.

"pink gloppy" . . . Rosenhan, *Odyssey into Lunacy,* chapter 7, 27.

"The accounting department has obviously" . . . Rosenhan, diary entry, undated.

"nameless terror" . . . Rosenhan, diary entry, undated.

"Distance permits us to control the terror" . . . Rosenhan, diary entry, undated.

"You got to talk to the doc" . . . Rosenhan, diary entry, undated.

"Drs. Exist to be conned" . . . Rosenhan, diary entry, undated.

"I might want to kill myself" . . . Rosenhan, diary entry, February 9, 1969.

"You've got to cooperate" . . . Rosenhan, diary entry, February 9, 1969.

"Have you got a moment, Mr. Harris?" . . .

Rosenhan, *Odyssey into Lunacy*, chapter 7, 17.

"in doing so behaved like a patient" . . . Rosenhan, *Odyssey into Lunacy*, chapter 7, 18.

"I then had the fantasy of kicking the door" . . . Rosenhan, diary entry, February 9, 1969.

"The blood rises" . . . Rosenhan, diary entry, February 10, 1969.

"There is none" . . . Rosenhan, diary entry, February 10, 1969.

"What are you writing?" . . . Rosenhan, diary entry, February 10, 1969.

I like you Mr. Harrison" . . . Rosenhan, diary entry, undated.

He began to break the walls . . . Rosenhan, *Odyssey into Lunacy,* chapter 7, 3–4.

In a case conference in 1967, a patient admitted . . . Zal, *Dancing with Medusa,* 50.

("in record time!") . . . Rosenhan, diary entry, February 12, 1969.

"couldn't tell many of the patients" . . . Rosenhan, diary entry, February 12, 1969.

"Nerves?" . . . Rosenhan, diary entry, undated.

"Look, this may be cold" . . . Rosenhan, diary entry, February 11, 1969.

"Feel like I'm leaving friends" . . . Rosenhan diary entry, February 14, 1969.

Myron Kaplan's note came from David Lurie's Haverford State Hospital medical records.

Stigma — in ancient Greece . . . Wulf Rossler, "The Stigma of Mental Disorders," *EMBO Reports* 17, no. 9 (2016), https://www.ncbi.nlm.nih.gov/pmc/articles/PMC5007563.

"A psychiatric label has a life" . . . Rosenhan, "On Being Sane in Insane Places," 253.

Tom Eagleton, a US senator . . . Ken Rudin, "The Eagleton Fiasco of 1972," NPR, March 7, 2007, https://www.npr.org/templates/story/story.php?storyId=7755888.

"quite shook" . . . Bea Patterson, phone interview, February 3, 2016.

Part Three

People ask, How did you get in there? . . . Susanna Kaysen, *Girl, Interrupted* (New York: Vintage Books, 1993), 5.

11: Getting In

I pulled together pseudopatients' stories with help from David's unpublished book,

Odyssey into Lunacy, scrap notes from his private files, and a spreadsheet titled "pseudo-patients," also from his private files.

Excerpt . . . Rosenhan, *Odyssey into Lunacy,* chapter 3, 15.

on March 29, 1969, at Rosenhan's lecture on altruism . . . The date and subject of his lecture at the Society for Research in Child Development (SRCD) were not explicitly stated in David's unpublished book. I was able to track them down thanks to help from Anne Purdue, director of operations at the SRCD, who found a copy of the 1969 event program.

"It was his thoughtfulness" . . . Rosenhan, *Odyssey into Lunacy,* chapter 3, 15.

"I should have been delighted" . . . Rosenhan, *Odyssey into Lunacy,* chapter 3, 16.

"John was particularly struck" . . . Rosenhan, *Odyssey into Lunacy,* chapter 3, 16.

"The procedure was simple" . . . Rosenhan, *Odyssey into Lunacy,* chapter 3, 17.

John called Rosenhan with news . . . The Beasleys' and Martha Coates's hospitalizations are discussed in various versions of chapters 3, 5, and 7 of *Odyssey into Lunacy.* Details about length of stay and hospital description are also found in David Rosenhan's unnamed pseudopatient list and a document titled "Hospital Descriptions" found in his private files.

"Bearded and burly" . . . Rosenhan, *Odyssey into Lunacy,* chapter 7, 31.

"I don't know what's troubling me" . . . Rosenhan, *Odyssey into Lunacy,* chapter 7, 41.

"I feel much better now" . . . Rosenhan, *Odyssey into Lunacy,* chapter 7, 43.

"evaluate their distress" . . . Rosenhan, *Odyssey into Lunacy,* chapter 3, 22.

"some wonderment about what" . . . Rosenhan, *Odyssey into Lunacy,* chapter 3, 20.

Laura Martin, the fifth pseudopatient . . . The Martins' hospitalizations are discussed in various versions of chapters 3, 5, 6, and 7 of *Odyssey into Lunacy.* Details about length of stay and hospital description are also found in David Rosenhan's unnamed pseudopatient list and a document titled "Hospital Descriptions" found in his private files.

"the top five [hospitals] in the country" . . . Rosenhan, "Hospital Descriptions," private files.

studies show that people with higher . . . Laeticia Eid, Katrina Heim, Sarah Doucette, Shannon McCloskey, Anne Duffy, and Paul Grof, "Bipolar Disorder and Socioeconomic Status: What Is the Nature of This Relationship?" *International Journal of Bipolar Disorder* 1, no. 9 (2013): 9, https://www.ncbi.nlm.nih.gov/pmc/articles/PMC4230315/.

"The hamburger was so coated" . . . Rosenhan, *Odyssey into Lunacy,* chapter 7, 37.

"We ourselves were seriously concerned" . . . Rosenhan, *Odyssey into Lunacy,* chapter 7, 39.

"With all due apologies for immodesty" . . . David Rosenhan, letter to Lorne M. Kendell, November 5, 1970, Correspondences Prior to 1974, Box 2, David L. Rosenhan Papers.

"There was further agreement" . . . George W. Goethals, letter to David Rosenhan, June 2, 1971, Correspondences Prior to 1974, Box 2, David L. Rosenhan Papers.

"The country is a hell of a lot more beautiful" . . . David Rosenhan, letter to Shel Feldman, July 28, 1970, Correspondences Prior to 1974, Box 2, David L. Rosenhan Papers.

"lucky we were to be here" . . . David Rosenhan, letter to Susan SantaMaria, July 30, 1970, Correspondences Prior to 1974, Box 2, David L. Rosenhan Papers.

misattributed to Mark Twain . . . David Mikkelson, "Mark Twain on Coldest

Winter," Snopes.com, https://www.snopes
.com/fact-check/and-never-the-twain-shall
-tweet.

"It was probably one" . . . Daryl Bem,
phone interview, April 13, 2016.

"one of the main motivations" . . . Rosenhan, *Odyssey into Lunacy,* chapter 3, 36.

"The ease with which we were able to gain
admission" . . . David Rosenhan, *Odyssey
into Lunacy,* chapter 3, 24.

how Rosenhan recruited Carl Wendt . . .
Carl Wendt's hospitalization (in some
places he is referred to as "Carl Wald,"
"Paul," and "Mark Schulz") is discussed
in various versions of chapters 3, 5, 6, 7,
and 8 of *Odyssey into Lunacy.* Details
about length of stay and hospital description are also found in David Rosenhan's
unnamed pseudopatient list and a document titled "Hospital Descriptions" found
in his private files.

"Much as it is common practice" . . .
Rosenhan, *Odyssey into Lunacy,* chapter
3, 29–30.

"What did you eat for breakfast?" . . .
Rosenhan, *Odyssey into Lunacy,* chapter
5, 8.

"I must be awfully tired" . . . Rosenhan,
Odyssey into Lunacy, chapter 7, 47.

"Bizarre as it may seem" . . . Rosenhan,

Odyssey into Lunacy, chapter 3, 32.

Excerpt of questionnaire from David Rosenhan private files.

Of 193 new patients . . . David Rosenhan,
"On Being Sane in Insane Places," 386.

"Were the patients sane or not?" . . . Sandra
Blakeslee, "8 Feign Insanity in Test and
Are Termed Insane," *New York Times,*
January 21, 1973, http://nyti.ms/1XVaRs9.

12: . . . And Only the Insane Knew Who Was Sane

Rosenhan then submitted his paper . . . David Rosenhan, letter to Phil Abelson,
August 14, 1972, private files. For more
on Phil Abelson's contribution to science
(and *Science*), see Jeremy Pearce, "Phil
Abelson, Chronicler of Scientific Advances, 91," *New York Times,* August 8,
2004, https://www.nytimes.com/2004/08/
08/us/philip-abelson-chronicler-of-sci
entific-advances-91.html.

"I read your article" . . . Letter to David
Rosenhan, Correspondences Prior to
1974, Box 8, David L. Rosenhan Papers.

"My name is Carl L. Harp" . . . Carl L.
Harp, letter to David Rosenhan, October
16, 1973, Correspondences Prior to 1974,
Box 8, David L. Rosenhan Papers.

"Dear Dr. David Rosenhan" . . . Letter to David Rosenhan, Correspondences Prior to 1974, Box 3, David L. Rosenhan Papers.

"I couldn't help but wonder" . . . David Rosenhan, letter, Correspondences Prior to 1974, Box 3, David L. Rosenhan Papers.

"I hope you forgive me" . . . David Rosenhan, letter to Pauline Lord, December 21, 1973, David L. Rosenhan Papers.

Los Angeles Times, ran it straight . . . George Alexander, "Eight Feign Insanity, Report on 12 Hospitals," *Los Angeles Times,* January 18, 1973: 1.

like the *Independent Record* in Helena . . . Sandra Vkajeskee, "Can Doctors Distinguish the Sane from the Insane?" *Independent Record,* January 28, 1973, 30.

The *Burlington Free Press* headlined its piece . . . Lee Hickling, " 'Mania,' 'Schizo' Labels Cause Wrangle," *Burlington Free Press,* November 7, 1975, 11.

The *Palm Beach Post* used . . . Sandra Blakeslee, ". . . And Only the Insane Knew Who Was Sane," *Palm Beach Post,* February 1, 1973, 17.

forced to sue him . . . *Doubleday & Company, Inc. v. David L. Rosenhan,* 5048/80, Supreme Court of the State of New York, County of New York, March 12, 1980.

"should not be permitted to testify" . . . Bruce J. Ennis and Thomas R. Litwick, "Psychiatry and the Presumption of Expertise: Flipping Coins in the Courtroom," *California Law Review* 62, no. 3 (1974).

judges increasingly overruled expert testimony . . . Paul S. Appelbaum, *Almost a Revolution: Mental Health Law and the Limits of Change* (Oxford: Oxford University Press, 1994).

"When the Rosenhan study was initiated" . . . Jeffrey Lieberman, phone interview, February 25, 2016.

"Rosenhan's study was akin to proving" . . . Robert Whitaker, *Mad in America: Bad Science, Bad Medicine, and the Enduring Mistreatment of the Mentally Ill* (New York: Basic Books, 2002), 170.

"It was a landmark study" . . . Allen Frances, phone interview, January 4, 2016.

"The most celebrated psychological experiment" . . . Michael E. Staub, *Madness Is Civilization: When the Diagnosis Was Social, 1948–1980* (Chicago: University of Chicago Press, 2011), 178.

Being gay then was considered a mental illness . . . Jack Drescher, "Out of DSM:

Depathologizing Homosexuality," *Behavioral Science* 5 (2015): 565–75.

there was a joke going around . . . Daryl Bem, phone interview, April 13, 2016.

sodomy between consenting adults, for example . . . Bingham, *Witness to the Revolution,* 180.

"Homosexuals are essentially disagreeable people" . . . Edmund Bergler, *Homosexuality: Disease or Way of Life?* (New York: Hill & Wang, 1956), 28–29.

"We can debate what is an illness" . . . *Before Stonewall* (documentary), directed by Greta Schiller and Robert Rosenberg, First Run Features, 1985.

"Homosexuality is in fact a mental illness" . . . "The Times They Are A-Changing," *The Sixties,* CNN.

Robert Galbraith Heath . . . For more on Robert Galbraith Heath, see Lone Frank, *The Pleasure Shock: The Rise of Deep Brain Stimulation and Its Forgotten Inventor* (New York: Dutton, 2018).

"continuous growing interest in women" . . . Cathy Gere, *Pain, Pleasure, and the Greater Good: From the Panopticon to the Skinner's Box and Beyond* (Chicago: University of Chicago Press, 2017), 193.

When news of the story . . . Gere, *Pain,*

Pleasure, and the Greater Good, 196–97.

"shrinked the headshrinkers" . . . Stuart Auerbach, "Gays and Dolls Battle the Shrinkers," *Washington Post,* May 15, 1970: 1.

"This lack of discipline is disgusting" . . . Ira Glass, "Episode 204: 81 Words," *This American Life,* National Public Radio, January 18, 2002, https://www.thisameri canlife.org/204/81-words.

"Psychiatry is the enemy incarnate" . . . "About This Document: Speech of 'Dr. Henry Anonymous' at the American Psychiatric Association 125th Annual Meeting, May 2, 1972," *Historical Society of Pennsylvania Digital Histories Project* (website), http://digitalhistory.hsp.org/pafrm/ doc/speech-dr-henry-anonymous-john -fryer-american-psychiatric-association -125th-annual-meeting.

One panelist was John Fryer . . . I relied on the following sources to depict John Fryer's Dr. Anonymous: Glass, "Episode 204: 81 Words"; John Fryer Papers at the Historical Society of Pennsylvania; and Dudley Clendinen, "John Fryer, 65, Psychiatrist Who Said He Was Gay in 1972, Dies," *New York Times,* March 5, 2003, http://www.nytimes.com/2003/03/05/ obituaries/05FRYE.html.

Fryer would cross paths with Rosenhan . . . I confirmed this detail with documents from both John Fryer's papers and David Rosenhan's private collection.

Handwritten excerpt of John Fryer's speech . . . John Fryer, "Speech for the American Psychiatric Association 125th Annual Meeting," undated, John Fryer Papers, Collection 3465, 1950–2000, Historical Society of Pennsylvania (Phila-delphia).

publicly reveal his identity as "Dr. Anony-mous" . . . Dudley Clendinen, "Dr. John Fryer, 65, Psychiatrist Who Said in 1972 He Was Gay," *New York Times,* March 5, 2003, https://www.nytimes.com/2003/03/05/us/dr-john-fryer-65-psychiatrist-who-said-in-1972-he-was-gay.html.

the APA's board of trustees . . . Decker, *The Making of the DSM-III,* 312.

"deep concerns over rampant criticism" . . . "Summary Report of the Special Policy Meeting of the Board of Trustees, Atlanta, Georgia. February 1–3, 1973," *American Journal of Psychiatry* 130, no. 6 (1973): 732.

"If you're going to have some people" . . . Jack Drescher, "An Interview with Rob-ert L. Spitzer," in Jack Drescher and Jo-seph P. Merlino, eds., *American Psychiatry and Homosexuality: An Oral History* (London: Routledge, 2007), 101.

a secret group . . . Kutchins and Kirk, *Making Us Crazy,* 69.

"Sexual Orientation Disturbance" . . . Drescher, "Out of DSM," 571.

described people distressed by their sexuality . . . For more information on gay rights and its connection to mental illness, see Eric Marcus, *Making Gay History: The Half-Century Fight for Lesbian and Gay Equal Rights* (New York: Harper Perennial, 2002).

A local newspaper satirized the removal . . . Vern L. Bullough, *Before Stonewall: Activists for Gay and Lesbian Rights in Historical Context* (London: Routledge, 2002), 249.

"Not only are women being punished" . . . Marcie Kaplan, "A Woman's View of the DSM-III," *American Psychologist* (July 1983): 791.

13: W. Underwood

the first paid installment . . . *Doubleday & Company, Inc. v. David L. Rosenhan.*

"More work of this kind" . . . Luther Nichols, letter to David Rosenhan, September 17, 1974, David Rosenhan private files.

Excerpt of David's outline . . . Outline for "Odyssey into Lunacy," David Rosenhan private files.

In 1973 and 1974, a Wilburn Underwood . . . Bill Underwood, Bert S. Moore, and David L. Rosenhan, "Affect and Self-Gratification," *Developmental Psychology* 8, no. 2 (1973): 209–14; and David L. Rosenhan, Bill Underwood, and Bert Moore, "Affect Moderates Self-Gratification and Altruism," *Journal of Personality and Social Psychology* 30, no. 4 (1974): 546–52.

Excerpt from yearbook . . . Stanford University, *Stanford Quad, 1973.* Print, Stanford University Archives.

Bert returned the email . . . Bert Moore, "Re: Request for help with contact information," email to Susannah Cahalan, January 15, 2015.

the soft-spoken, red-bearded . . . David Rosenhan, *Odyssey into Lunacy,* chapter 3, 38.

Hi Susannah . . . Bill Underwood, "Re: Request for Interview," email to Susannah Cahalan, January 31, 2015.

14: Crazy Eights

The bulk of this chapter came from various interviews with Bill Underwood and Maryon Underwood over a four-year period, but especially the first time I visited them in person at their home in Texas, February 9, 2015.

Charles Whitman climbed the tower . . . For more on Charles Whitman, see the powerful documentary *Tower,* directed by Keith Maitland, Go-Valley Productions, 2016.

"I don't really understand myself" . . . Lauren Silverman, "Gun Violence and Mental Health Laws, 50 Years After Texas Tower Sniper," *Morning Edition,* National Public Radio, July 29, 2016, https://www.npr.org/sections/health-shots/2016/07/29/487767127/gun-violence-and-mental-health-laws-50-years-after-texas-tower-sniper.

An autopsy revealed a glioblastoma . . . David Eagleman, "The Brain on Trial," *The Atlantic,* July–August 2011, https://www.theatlantic.com/magazine/archive/2011/07/the-brain-on-trial/308520.

A flurry of brain studies followed . . . Thank you to Dr. William Carpenter for insights into the history of neuro-imaging.

enlarged ventricles . . . N. C. Andreasen,

S. A. Olsen, J. W. Dennert, and M. R. Smith, "Ventricular Enlargement in Schizophrenia: Relationship to Positive and Negative Symptoms," *American Journal of Psychiatry* 139, no. 3 (1982): 297–302.

gray matter thinning . . . Martha E. Shenton, Chandlee C. Dickey, Melissa Frumin, and Robert W. McCarley, "A Review of MRI Findings in Schizophrenia," *Schizophrenia Research* 49, nos. 1–2 (2001): 1–52. Thank you to Dr. William Carpenter for speaking with me about the advancements and continued limitations of scanning technology.

But the hope that CT scans . . . Robin Murray, "Mistakes I Have Made in My Research Career," *Schizophrenia Bulletin* 43, no. 1 (2017): 253–56, https://academic .oup.com/schizophreniabulletin/article/43/ 2/253/2730504.

the "riddle of schizophrenia" . . . Nancy Andreasen, *The Broken Brain: The Biological Revolution in Psychiatry* (New York: Harper & Row, 1984), 53.

Everything from sustained antipsychotic use . . . Thank you to Maree Webster, who, on January 14, 2016, explained many of the complexities of studying the brain and also showed me around the truly jaw-dropping brain bank that she runs.

"despite vigorous study over the past century" . . . R. Tandon, M. S. Keshavan, and H. A. Nasrallah, "Schizophrenia, Just the Facts: What We Know in 2008. Part 1: Overview," *Schizophrenia Research* 100 (2008): 4, 11.

Stanford University campus's Bing Nursery . . . Janine Zacharia, "The Bing 'Marshmallow Studies': 50 Years of Continuing Research," Distinguished Lecture Series, Stanford, https://bingschool .stanford.edu/news/bing-marshmallow -studies-50-years-continuing-research.

Mischel found that a child's ability . . . W. Mischel et al., "Delay of Gratification in Children," *Science* 24, no. 4 (1989): 933–38.

All Robyn remembers . . . Robyn Harrigan, phone interview, November 2, 2016.

"I didn't want David to be my lifeline" . . . Craig Haney, in-person interview, February 17, 2017.

"least likely" to be admitted . . . Rosenhan, *Odyssey into Lunacy,* chapter 3, 38.

the Lanterman-Petris-Short Act . . . Marc F. Abramson, "The Criminalization of Mentally Disordered Behavior: Possible Side-Effect of a New Mental Health Law," *Hospital & Community Psychiatry* 23, no. 4

(1972): 101–5.

Located less than half an hour . . . The history of Agnews was compiled from a variety of sources, including from a private tour of the Agnews Museum provided to me by Kathleen Lee on October 21, 2015. Santa Clara University's archives were also helpful: "Agnews State Hospital," Silicon Valley History online, Santa Clara University Library Digital Collections, http://content.scu.edu/cdm/landingpage/collection/svhocdm.

"They were tense times" . . . Izzy Talesnick, in-person interview, October 22, 2015.

case #115733 . . . I found this in Bill Underwood's Agnews State Hospital medical records, tracked down with help from Bill Underwood and Florence Keller.

I had tracked down the ACLU lawyer . . . Robert Bartels, phone interview, January 15, 2015.

15: Ward 11

on a special unit called Ward 11 . . . Alma Menn, in-person interview, October 23, 2015. I've also seen it referred to as I-Ward.

"Not only do people publicly neck" . . . Jane

Howard, "Inhibitions Thrown to the Gentle Winds: A New Movement to Unlock the Potential of What People Could Be — But Aren't," *Life,* July 12, 1968, 48–65.

Bob Dylan visited . . . Art Harris, "Esalen: From '60s Outpost to the Me Generation," *Washington Post,* September 24, 1978, https://www.washingtonpost.com/archive/opinions/1978/09/24/esalen-from-60s-outpost-to-the-me-generation/f1db58bb-e77f-4bdf-9457-e07e6b4cc800/?utm_term=.a8248c047098.

Charles Manson showed up . . . Walter Truett Anderson, *The Upstart Spring: Esalen and the American Awakening* (Boston: Addison-Wesley, 1983), 239.

Dick Price was supposed to follow . . . Dick Price's backstory was compiled from a variety of sources, including Jeffrey J. Kripal, *Esalen: America and the Religion of No Religion* (Chicago: University of Chicago Press, 2007); Wade Hudson, "Dick Price: An Interview," Esalen.org, 1985, https://www.esalen.org/page/dick-price-interview; and Anderson, *The Upstart Spring.*

he heard a disembodied voice . . . Anderson, *The Upstart Spring,* 38.

"He felt a tremendous opening up" . . . Anderson, *The Upstart Spring,* 39.

a fancier private hospital . . . The description of the Institute of Living came from an in-person tour of the museum on their grounds; also from Luke Dittrich, *Patient H.M.: A Story of Memory, Madness, and Family Secrets* (New York: Random House, 2016), 60.

The Chatterbox, which once ran an illustration . . . Barry Werth, "Father's Helper," *New Yorker,* June 9, 2003, https://www .newyorker.com/magazine/2003/06/09/ fathers-helper.

The institute's psychiatrist-in-chief, Dr. Francis J. Braceland . . . Werth, "Father's Helper."

"private prison" . . . The Gestalt Legacy Project, *The Life and Practice of Richard Price: A Gestalt Biography* (Morrisville, NC: Lulu Press, 2017), 39.

he underwent ten electroshock therapies . . . Kripal, *Esalen,* 80.

"the complete debilitator" . . . The Gestalt Legacy Project, *The Life and Practice of Richard Price,* 40.

This is what Dick would have faced . . . The description of insulin coma therapy came from "A Brilliant Madness," *American Experience,* PBS, directed by Mark Samels, WGBH Educational Foundation, 2002.

he underwent fifty-nine of these therapies . . . The Gestalt Legacy Project, *The
Life and Practice of Richard Price,* 4.

put on over seventy pounds . . . The Gestalt
Legacy Project, *The Life and Practice of
Richard Price,* 40.

screen actress Gene Tierney . . . Kent De-
maret, "Gene Tierney Began Her Trip
Back from Madness on a Ledge 14 Floors
Above the Street," *People,* May, 7, 1979,
https://people.com/archive/gene-tierney
-began-her-trip-back-from-madness-on-a
-ledge-14-floors-above-the-street-vol-11
-no-18.

"would serve people" . . . The Gestalt
Legacy Project, *The Life and Practice of
Richard Price,* 77.

"live through experience" . . . Hudson,
"Dick Price: An Interview."

R. D. Laing came to Esalen . . . For more
on Kingsley Hall, see the documentary
Asylum, directed by Peter Robinson, 1972.
Thank you to Richard Adams, one of the
cameramen who filmed the movie, who
gave me valuable insights and supplied an
unedited version.

That same year psychologist Julian Silverman . . . Kripal, *Esalen,* 169.

befriended the Grateful Dead . . . Alma

Menn, in-person interview, October 23, 2015.

John Rosen, the inventor of "direct analysis" . . . Joel Paris, *Fall of an Icon: Psychoanalysis and Academic Psychiatry* (Toronto: University of Toronto Press, 2005), 30.

Rosen later lost his license . . . United Press International, "79-Year-Old Former Doctor Loses License to Practice," *Logansport Pharos-Tribune,* April 8, 1983, 3.

"ding dong city" . . . The Gestalt Legacy Project, *The Life and Practice of Richard Price,* 76.

They selected a few Agnews staff . . . My description of Ward 11 came from a variety of sources, including interviews with Alma Menn (October 23, 2015) and Voyce Hendrix (December 8, 2016); the research paper published on it: Maurice Rappaport et al., "Are There Schizophrenics for Whom Drugs May Be Unnecessary or Contraindicated?" *International Pharmapsychiatry* 13 (1978): 100–111; and secondary sources like Michael Cornwall, "The Esalen Connection: Fifty Years of Re-Visioning Madness and Trying to Transform the World," *Mad in America* (blog), December 12, 2013, https://www

.madinamerica.com/2013/12/esalen -connection-fifty-years-re-visioning -madness-trying-transform-world.

"The first thing we did" . . . Alma Menn, in-person interview, October 23, 2015.

published in the 1978 paper . . . Rappaport, "Are There Schizophrenics."

a series of "med-free sanctuaries" . . . Michael Cornwall, "Remembering a Medication-Free Madness Sanctuary," *Mad in America* (blog), February 3, 2012, https://www.madinamerica.com/2012/02/ remembering-a-medication-free-madness -sanctuary.

Soteria House, an experiment in communal living . . . John R. Bola and Loren Mosher, "Treatment of Acute Psychosis Without Neuroleptics: Two-Year Outcomes from Soteria Project," *Journal of Nervous Disease* 191, no. 4 (2003): 219–29.

The average stay was forty-two days . . . John Reed and Richard Bentall, eds., *Models of Madness: Psychological, Social, and Biological Approaches to Schizophrenia* (London: Routledge, 2004), 358.

three to five times lower . . . Reed and Bentall, *Models of Madness,* 358.

One former Soteria resident . . . B. Mooney, phone interview, January 18, 2017.

in the clubhouse model . . . For more on

the clubhouse model approach, see Colleen McKay, Katie L. Nugent, Matthew Johnsen, William W. Easton, and Charles W. Lidz, "A Systematic Review of Evidence for the Clubhouse Model of Psychosocial Rehabilitation," *Administration and Policy in Mental Health and Mental Health Services* 45, no. 1 (2018): 28–47, https://www.ncbi.nlm.nih.gov/pubmed/27580614.

We see it also in Geel . . . For more on Geel, a fascinating place with an even more fascinating history, see Angus Chen, "For Centuries, a Small Town Has Embraced Strangers with Mental Illness," *NPR,* July 1, 2016, https://www.npr.org/sections/health-shots/2016/07/01/484083305/for-centuries-a-small-town-has-embraced-strangers-with-mental-illness.

In Trieste, Italy . . . Elena Portacolone, Steven P. Segal, Roberto Mezzina, and Nancy Scheper-Hughes, "A Tale of Two Cities: The Exploration of the Trieste Public Psychiatry Model in San Francisco," *Culture, Medicine, and Psychiatry* 39, no. 4 (2015). Thank you also to Kerry Morrison for making me aware of this amazing place.

Price suffered another break . . . The Gestalt Legacy Project, *The Life and Practice of*

Richard Price, 83.

16: *Soul on Ice*

This chapter again was based on several in-person and phone interviews with the Underwoods.

nicknamed "Dr. Sparky" . . . Izzy Talesnick and Jo Gampon, in-person interview, October 22, 2015.

Ugo Cerletti, who came up with the idea . . . Valenstein, *Great and Desperate Cures,* 51.

A psych technician from that era . . . Interview with "Jim" at Agnews Historic Cemetery and Museum, October 21, 2015, http://santaclaraca.gov/Home/Components/ServiceDirectory/ServiceDirectory/1316/2674.

I saw an electroshock box . . . Interview with Anthony Ortega at Patton Hospital Museum, October 29, 2016, http://www.dsh .ca.gov/Patton/Museum.aspx.

when Olivia de Havilland seizes . . . *The Snake Pit* (film), directed by Anatole Litvak, Twentieth Century Fox Film Corporation, 1948.

Patients would sometimes break their backs . . . Valenstein, *Great and Desperate Cures,* 53.

"clever little procedure" . . . Kesey, *One*

Flew Over the Cuckoo's Nest, 62.

patients who are "treatment resistant" . . . S. G. Korenstein and R. K. Schneider, "Clinical Features of Treatment-Resistant Depression," *Journal of Clinical Psychiatry* 62, no. 16 (2001): 18–25.

"now a fully safe and painless procedure" . . . Charles Kellner, "ECT Today: The Good It Can Do," *Psychiatric Times,* September 15, 2010, http://www .psychiatrictimes.com/electroconvulsive -therapy/ect-today-good-it-can-do.

is paired with an immobilizing agent . . . Scott O. Lilienfeld, "The Truth About Shock Therapy," *Scientific American,* May 1, 2014, https://www.scientificamerican .com/article/the-truth-about-shock -therapy.

In one study, 65 percent of patients . . . Hilary J. Bernstein et al., "Patient Attitudes About ECT After Treatment," *Psychiatric Annals* 28 (1998): 524–27, https://www .healio.com/psychiatry/journals/psycann/ 1998-9-28-9/%7B189440aa-c05e-4cbb -ae9b-992c9ec85dba%7D/patient-attitu des-about-ect-after-treatment. For a hilarious pro-ECT take, see Carrie Fisher, *Shockaholic* (New York: Simon & Schuster, 2011).

"a crime against humanity" . . . "Resolution

Against Electroshock: A Crime Against Humanity," ECT.org, http://www.ect.org/resources/resolution.html.

more hospitals have used it on the East Coast . . . Brady G. Case, David N. Bertolio, Eugene M. Laska, Lawrence H. Price, Carole E. Siegel, Mark Olfson, and Steven C. Marcus, "Declining Use of Electroconvulsive Therapy in US General Hospitals," *Biological Psychiatry* 73, no. 2 (2013): 119–26.

Hollywood's vilification of the procedure . . . Garry Walterand Andrew McDonald, "About to Have ECT? Fine, But Don't Watch It in the Movies," *Psychiatric Times,* June 1, 2004, https://www.psychiatrictimes.com/antisocial-personality-disorder/about-have-ect-fine-dont-watch-it-movies-sorry-portrayal-ect-film/page/0/1.

"reason for discharge" blank . . . Special thanks to Bill Underwood and Florence Keller for tracking down this record.

ten days less than the norm . . . Scull, *Decarceration,* 147.

which hovered around 130 days . . . Scull, *Decarceration,* 147.

In 2009, Agnews closed for good . . . Linda Goldston, "After More than 120 Years, Agnews Is Closing This Week," *Mercury News,* March 24, 2009, https://www

.mercurynews.com/2009/03/24/after-more
-than-120-years-agnews-is-closing-this
-week.

17: Rosemary Kennedy

This chapter was aided tremendously by the work of E. Fuller Torrey in his book *American Psychosis: How the Federal Government Destroyed the Mental Illness Treatment System* (Oxford: Oxford University Press, 2013), as well as several in-person and phone interviews conducted with him.

"The anti-psychiatrists could now" . . . Rael Jean Isaac and Virginia Armat, *Madness in the Streets: How Psychiatry and the Law Abandoned the Mentally Ill* (Arlington, VA: Treatment Advocacy Center, 1990), 56.

these hospitals were "superfluous" institutions . . . Scull, *Decarceration,* 73.

"therapeutic tyranny" . . . Thomas Szasz, *The Manufacture of Madness: A Comparative Study of the Inquisition and the Mental Health Movement* (Syracuse: Syracuse University Press, 1970).

"merely a symptom of an outdated system" . . . George S. Stevenson, "Needed: A Plan for the Mentally Ill," *New York Times,* July 27, 1947.

"liquidated as rapidly" . . . Isaac and Armat, *Madness in the Streets,* 69.

California governor Ronald Reagan closed . . . Torrey, *American Psychosis.*

Modesto . . . "Inventory of the Department of Mental Hygiene — Modesto State Hospital Records," *Online Archive of California,* https://oac.cdlib.org/findaid/ark:/13030/tf267n98b9/?query=Modesto.

Dewitt . . . "Inventory of the Department of Mental Hygiene — Dewitt State Hospital Records," *Online Archive of California,* https://oac.cdlib.org/findaid/ark:/13030/tf396n990k/?query=Dewitt+state+hospital.

and Mendocino State Hospitals . . . "Inventory of the Department of Mental Hygiene — Mendocino State Hospital Records," *Online Archive of California,* https://oac.cdlib.org/findaid/ark:/13030/tf2c6001q2/.

converted Agnews into an institution. . . . "Agnews Developmental Center," *State of California Department of Developmental Services,* https://www.dds.ca.gov/Agnews/.

"better off outside of a hospital" . . . E. Fuller Torrey, *Out of the Shadows: Confronting America's Mental Illness Crisis* (New York: Wiley, 1996), 143.

Rosemary Kennedy's first hours . . . Rosemary's story was compiled from two recent biographies: Kate Clifford Larson,

Rosemary: The Hidden Kennedy Daughter (New York: Houghton Mifflin Harcourt, 2015); and Elizabeth Koehler-Pentacoff, *Missing Kennedy: Rosemary Kennedy and the Secret Bonds of Four Women* (Baltimore: Bancroft Press, 2015).

the official label was "mentally retarded" . . . Larson, *Rosemary,* 45.

Moniz, who received a Nobel Prize . . . For more about António Egas Moniz and Walter Freeman, read Jack El-Hai, *The Lobotomist: A Maverick Medical Genius and His Tragic Quest to Rid the World of Mental Illness* (Hoboken, NJ: Wiley, 2005).

Neurologist Freeman would adapt . . . For a devastating, must-read piece on Walter Freeman's legacy, see Michael M. Phillips, "The Lobotomy Files: One Doctor's Legacy," *Wall Street Journal,* December 13, 2013, http://projects.wsj.com/lobotomyfiles/?ch=two.

Sixty percent of lobotomies were conducted on women . . . Jack El-Hai, "Race and Gender in the Selection of Patients for Lobotomy," *Wonders & Marvels,* http://www.wondersandmarvels.com/2016/12/race-gender-selection-patients-lobotomy.html.

one study in Europe found that 84 percent . . . Louis-Marie Terrier, Marc

Leveque, and Aymeric Amelot, "Most Lobotomies Were Done on Women" (letter to the editor), *Nature* 548 (2017): 523.

"It's nothing we want done" . . . Lyz Lenz, "The Secret Lobotomy of Rosemary Kennedy," *Marie Claire,* March 31, 2017, https://www.marieclaire.com/celebrity/a2 6261/secret-lobotomy-rosemary-kennedy.

Dr. Watts drilled burr holes . . . Dittrich, *Patient H.M.,* 75–77, and Larson, *Rosemary,* 168–70.

"a painting that had been brutally slashed" . . . Laurence Leamer, *The Kennedy Women: The Saga of an American Family* (New York: Random House, 1995), 338.

she didn't visit her daughter . . . Larson, *Rosemary,* 175.

where she remained until her death . . . "Rosemary Kennedy, Senator's Sister, 86, Dies," *New York Times,* January 8, 2005, https://www.nytimes.com/2005/01/08/obituaries/rosemary-kennedy-senators-sister-86-dies.html.

"yet more danger, death" . . . Larson, *Rosemary,* 180.

"I have sent to the Congress today" . . . John F. Kennedy, "Remarks upon Signing a Bill for the Construction of Mental Retardation Facilities and Community

Mental Health Centers, 31 October 1963," John F. Kennedy Presidential Library and Museum archives, https://www.jfklibrary.org/asset-viewer/archives/JFKWHA/1963/JFKWHA-161-007/JFKWHA-161-007.

"U.S. Army psychiatrists in World War II" . . . Appelbaum, *Almost a Revolution,* 8.

"prolonged hospital stays might" . . . Appelbaum, *Almost a Revolution,* 8.

"an ongoing exodus of biblical proportions" . . . Torrey, *American Psychosis,* 76.

"payer, insurer, and regulator" . . . Richard G. Frank, "The Creation of Medicare and Medicaid: The Emergence of Insurance and Markets for Mental Health Services," *Psychiatric Services* 51, no. 4 (2000): 467.

Institutions for Mental Diseases (IMD) exclusion . . . "The Medicaid IMD Exclusion: An Overview and Opportunities for Reform," Legal Action Center, https://lac.org/wp-content/uploads/2014/07/IMD_exclusion_fact_sheet.pdf.

leaving the mentally ill to vie . . . Torrey, *American Psychosis,* 164.

Medicaid continues to be the United States' . . . Alisa Roth, *Insane: America's Criminal Treatment of Mental Illness* (New

York: Basic Books, 2018), 91.

" 'medicalized' treatment settings" . . . Frank, "The Creation of Medicare and Medicaid," 467.

federal mental health parity law . . . For more on the Mental Health Parity and Addiction Equity Act (MHPAEA), see https://www.cms.gov/cciio/programs-and -initiatives/other-insurance-protections/ mhpaea_factsheet.html.

insurance companies now reimburse . . . Lizzie O'Leary and Peter Balonon-Rosen, "When It Comes to Insurance Money, Mental Health Is Not Treated Equal," *Marketplace,* January 5, 2018, https://www .marketplace.org/2018/01/05/health-care/ doctors-get-more-insurance-money -psychiatrists-when-treating-mental -health.

just over half of psychiatrists take insurance . . . Tara F. Bishop, Matthew J. Press, Salomeh Keyhani, and Harold Alan Pincus, "Acceptance of Insurance by Psychiatrists and the Implications for Access to Mental Health Care," *JAMA Psychiatry* 71, no. 2 (2014): 176–81, https://www.ncbi .nlm.nih.gov/pmc/articles/PMC3967759.

A series of landmark acts . . . For a great treatment of the landmark rulings that changed health policy, see Appelbaum, *Al-*

most a Revolution.

"mental illness treatment system had been essentially beheaded" . . . Torrey, *American Psychosis,* 89.

dropped by almost 50 percent . . . Scull, *Decarceration,* 68.

another 50 percent to 132,164 . . . David Mechanic, *Inescapable Decisions: The Imperative of Health Reform* (Piscataway, NJ: Transaction Publishers, 1994), 172.

Today 90 percent of the beds . . . This percentage change comes from comparing the number of beds in JFK's era (504,600) to 52,539 in 2004, found in E. Fuller Torrey et al., "The Shortage of Public Hospital Beds for Mentally Ill Persons: A Report of the Treatment Advocacy Center," Treatment Advocacy Center, Arlington, VA, https://www.treatmentadvocacycenter.org/storage/documents/the_shortage_of_publichospital_beds.pdf.

"small long-term state hospital wards" . . . H. Richard Lamb and Victor Goertzel, "Discharged Mental Patients — Are They Really in the Community?" *Archives of General Psychiatry* 24, no. 1 (1971): 29–34.

"We could see the light" . . . Dominique Kinney, in-person interview, October 29, 2016.

Part Four

When the going gets weird . . . Hunter S. Thompson, "Fear and Loathing at the Super Bowl," *Rolling Stone,* February 28, 1974, https://www.rollingstone.com/culture/culture-sports/fear-and-loathing-at-the-super-bowl-37345/.

18: The Truth Seeker

"I'm simply not sure that more money" . . . David Rosenhan, letter to James Floyd, January 24, 1973, David L. Rosenhan Papers.

"Bill Dixon's" hospital held 8,000 patients . . . Rosenhan, pseudopatient list.

Rosenhan wrote that all the pseudopatients . . . Rosenhan, "On Being Sane in Insane Places," 252.

71 percent of psychiatrists moved on . . . Rosenhan, "On Being Sane in Insane Places," Early Undated Draft, private files.

"He certainly wouldn't have gotten" . . . Bill Underwood, email to Susannah Cahalan, March 26, 2017.

"Seriously flawed by methodological inadequacies" . . . Fleischman, letter to the editor, 356.

"It appears that the pseudopatient gathered" . . . Thaler, letter to the editor, 358.

"If I were to drink a quart of blood" . . . Seymour S. Kety, "From Rationalization to Reason," *American Journal of Psychiatry* 131 (1974): 959.

"To point out that Rosenhan's conclusion" . . . J. Vance Israel, letter to the editor, *Science,* April 27, 1973, 358.

A representative said that she . . . Meagan Phelan, email to Susannah Cahalan, March 14, 2016. The message read: "Thank you for your query. Unfortunately the peer review process of research articles like the one you cite below is confidential, so I'm afraid I cannot provide answers to your questions."

"mainly because they have" . . . David Rosenhan, letter to Henry O. Patterson, July 31, 1975, David L. Rosenhan Papers.

"Submitting to *Science* [may have been] a trick" . . . Ben Harris, phone interview, December 19, 2016.

"Some foods taste delicious" . . . Robert Spitzer, "On Pseudoscience in Science, Logic in Remission, and Psychiatric Diagnosis: A Critique of Rosenhan's 'On Being Sane in Insane Places,' " *Journal of Abnormal Psychology* 84, no. 5 (1975): 442–52.

"Sane comes closest to what" . . . David

Rosenhan, letter to Alexander Nies, July 10, 1973, David L. Rosenhan Papers.

"Until now, I have assumed" . . . Spitzer, "On Pseudoscience in Science," 447.

The first letter opened "Dear Dave" . . . Robert Spitzer, letter to David Rosenhan, December 5, 1974, David L. Rosenhan Papers.

A close reading of Rosenhan's response . . . David Rosenhan, letter to Robert Spitzer, January 15, 1975, David L. Rosenhan Papers.

Spitzer himself had been long obsessed . . . Alix Spiegel, "The Dictionary of Disorder," *New Yorker,* January 3, 2005, https:// www.newyorker.com/magazine/2005/01/ 03/the-dictionary-of-disorder.

Reichian psychology and its orgone box therapy . . . Decker, *The Making of the DSM-III,* 89.

Another Reichian with a rumored orgone box . . . Tim Murphy, " 'You Might Very Well Be the Cause of Cancer': Read Bernie Sanders' 1970s-Era Essays," *Mother Jones,* July 6, 2015, https://www.mother jones.com/politics/2015/07/bernie-sanders -vermont-freeman-sexual-freedom-fluor ide.

Spitzer's grandfather had pitched his own wheelchair . . . Janet Williams, phone interview, May 27, 2017; email confirmation with his two children, Laura and Daniel Spitzer.

His mother struggled with depression . . . Janet Williams, phone interview, March 16, 2016.

He struggled with depression . . . Janet Williams, phone interview, March, 16, 2016.

"a truth seeker" . . . Janet Williams, phone interview, April 27, 2017.

"[This] implies that I have something to conceal" . . . Rosenhan, letter to Spitzer, January 15, 1975.

"Let me make clear" . . . David Rosenhan, letter to the editor, *Science,* April 27, 1973, 369.

"Perhaps all that we can hope for" . . . Spitzer, letter to Rosenhan, March 5, 1975.

"You now have it from myself and the superintendent" . . . Rosenhan, letter to Spitzer, January 15, 1975.

"You're not crazy" . . . Rosenhan, "On Being Sane in Insane Places," 385.

19: "All Other Questions Follow from That"

"no further alterations" . . . Rosenhan, "On Being Sane in Insane Places," 383.

This is what Dr. Bartlett recorded . . . Excerpt from Haverford State Hospital medical records.

Hallucinations and disturbances in thought patterns . . . "Schizophrenia: Symptoms and Causes," Mayo Clinic, https://www.mayoclinic.org/diseases-conditions/schizophrenia/symptoms-causes/syc-20354443.

"thought broadcasting," or the belief . . . Theodore A. Stern, *Massachusetts General Hospital Handbook of General Hospital Psychiatry* (Philadelphia: Saunders, 2010), 531.

an "existential permeability" . . . Clara Kean, "Battling with the Life Instinct: The Paradox of the Self and Suicidal Behavior in Psychosis," *Schizophrenia Bulletin* 37, no. 1 (2011): 4–7, https://academic.oup.com/schizophreniabulletin/article/37/1/4/1932702; and Clara Kean, "Silencing the Self: Schizophrenia as Self-Disturbance," *Schizophrenia Bulletin* 35, no. 6 (2009): 1034–36, https://www.ncbi.nlm.nih.gov/pmc/articles/PMC2762621/.

Rosenhan "dated his illness to *ten years ago*" . . . Haverford State Hospital medical records.

"much clearer picture of schizophrenia" . . . Dr. Michael Meade, email to Susannah

Cahalan, March 17, 2019.

Medical record excerpt from David Lurie's Haverford State Hospital medical records.

"Active psychosis is one" . . . Meade, email to Cahalan.

"It seems to me that any sentient human being" . . . Florence Keller, email to Susannah Cahalan, November 9, 2017.

"placed the bottom of a copper pot" . . . Haverford State Hospital medical records.

Dr. Frank "Lewis" Bartlett had died . . . "Services Pending for Psychiatrist F. Lewis Bartlett," TulsaWorld.com, May 26, 1989, https://www.tulsaworld.com/archives/services-pending-for-psychiatrist-f-lewis-bartlett/article_01472847-cb55-5e2e-b8b2-9daaf6c4f704.html.

Dr. Bartlett's interest in psychiatry . . . Information about Dr. Bartlett's life came via interviews with Mary Bartlett, Claudia Bushee, and Carole Adrienne Murphy.

coined the term *institutional peonage* . . . F. Lewis Bartlett, "Institutional Peonage: Our Exploitation of Mental Patients," *Atlantic Monthly,* July 1964, 116–18.

gave him a "creepy feeling" . . . F. Lewis Bartlett, letter to Ken Kesey, March 16, 1962, Mary Bartlett personal files.

"I just have this picture of Lew" . . . Mary

Bartlett, phone interview, January 30, 2017.

And then there was the interview . . . Ervin Staub, phone interview, August 25, 2017.

Medical record photo of David . . . Excerpt from David Lurie's Haverford State Hospital medical records.

Excerpt from David Lurie's Haverford State Hospital medical records.

Excerpt from David Rosenhan, "On Being Sane in Insane Places," 387.

"The facts of the case were unintentionally distorted" . . . Rosenhan, "On Being Sane in Insane Places," 387.

20: Criterionating

I am so grateful to the work of Hannah Decker and her engrossing and informative book *The Making of the DSM-III* for help in writing this chapter. Thanks also to Janet Williams, Michael First, Allen Frances, and Ken Kendler for providing some firsthand insight into the process.

how Spitzer managed to get his hands . . . The information about how "David Lurie's" medical records ended up in Spitzer's hands came from the letters exchanged between Rosenhan and Spitzer.

denouncement of his prior research . . .

Robert Spitzer (guest), "Spitzer's Apology Changes 'Ex-Gay' Debate," *Talk of the Nation,* National Public Radio, May 21, 2012, https://www.npr.org/2012/05/21/153213796/spitzers-apology-changes-ex-gay-debate.

"Dr. Robert L. Spitzer, who gave psychiatry" . . . Benedict Carey, "Robert Spitzer, 83, Dies; Psychiatrist Set Rigorous Standards for Diagnosis," *New York Times,* December 26, 2015, https://www.nytimes.com/2015/12/27/us/robert-spitzer-psychiatrist-who-set-rigorous-standards-for-diagnosis-dies-at-83.html.

"the best thing I have ever written" . . . Decker, *The Making of the DSM-III,* 103.

writing a follow-up on Rosenhan's study . . . Robert Spitzer, "More on Pseudoscience in Science and the Case for Psychiatric Diagnosis," *Archives of General Psychiatry* 33, no. 4 (1976): 466, https://jamanetwork.com/journals/jamapsychiatry/article-abstract/491528?resultClick=1.

"For Spitzer, paradoxically, Rosenhan's study" . . . Scull, *Psychiatry and Its Discontents,* 282.

"fateful point in the history" . . . Gerald L. Klerman, "The Advantages of *DSM-III,*" *American Journal of Psychiatry* 141, no. 4 (1984): 539.

"They were determined to create" . . . Luhrmann, *Of Two Minds,* 225.

The Wash U group also referred to themselves . . . Decker, *The Making of the DSM-III,* 115.

whose "guns [were] pointed" at psychoanalysis . . . Decker, *The Making of the DSM-III,* 225.

they kept a picture of Freud . . . Decker, *The Making of the DSM-III,* 71.

the "Feighner Criteria" . . . John P. Feighner, Eli Robins, Samuel B. Guze, Robert A. Woodruff, George Winokur, and Rodrigo Munoz, "Diagnostic Criteria for Use in Psychiatric Research," *Archives of General Psychiatry* 26 (January 1972): 57–63.

494 pages, compared with the *DSM-II* . . . Rick Mayes and Allan V. Horwitz, "DSM-III and the Revolution in the Classification of Mental Illness," *Journal of the History of the Behavioral Sciences* 41, no. 3 (2005): 25.

The *DSM-III* introduced "axes" . . . American Psychiatric Association, *Diagnostic and Statistical Manual of Mental Disorders,* 3rd ed. (Washington, DC: American Psychiatric Association, 1980).

"conditions and patterns of behavior" . . .

Kutchins and Kirk, *Making Us Crazy,* 176.

"It is as important to psychiatrists" . . . Gary Greenberg, "Inside the Battle to Define Mental Illness," *Wired,* December 27, 2010, https://www.wired.com/2010/12/ff_dsmv.

creating "rich pickings" . . . Healy, *The Antidepressant Era,* 213.

"is conceptualized as a clinically significant" . . . American Psychiatric Association, *Diagnostic and Statistical Manual of Mental Disorders,* 3rd ed., 6.

"based on the tradition of separating these disorders" . . . American Psychiatric Association, *Diagnostic and Statistical Manual of Mental Disorders,* 3rd ed., 8.

"Hence, this manual uses" . . . American Psychiatric Association, *Diagnostic and Statistical Manual of Mental Disorders,* 3rd ed., 8.

also known as the field's remedicalization . . . Wilson, "DSM-III and the Transformation of American Psychiatry," 399.

Gerald Klerman called it "a victory" . . . Shorter, *A History of Psychiatry,* 302.

"no longer must carry the burden" . . . Andreasen, *The Broken Brain,* 249.

"When we would write a criterion" . . . Janet Williams, phone interview, May 27, 2017.

"Rosenhan's pseudopatients would never" . . . Luhrmann, *Of Two Minds,* 231.

"What Bob [Spitzer] did" . . . Allen Frances, phone interview, January 4, 2016.

21: The SCID

his memorial lecture . . . The Robert L. Spitzer Memorial Lecture took place on October 26, 2016, at Columbia's Herbert Pardes Building.

Rosenhan included this interaction . . . David Rosenhan, "On Being Sane in Insane Places," 255.

"The following year David Rosenhan published" . . . Michael First, Spitzer Memorial Lecture, October 26, 2016.

"reification of psychiatric diagnoses" . . . Ken Kendler, Spitzer Memorial Lecture, October 26, 2016.

"Rather than heading off" . . . Shorter, *A History of Psychiatry,* 302.

"clustered around Spitzer". . . . Decker, *The Making of the DSM-III,* 109.

seventy to eighty hours a week . . . Janet Williams, phone interview, May 27, 2017.

"There would be these meetings" . . . Spiegel, "The Dictionary of Disorder."

"There was very little systematic research" . . . https://www.newyorker.com/

magazine/2005/01/03/ the-dictionary-of-disorder Spiegel, "The Dictionary of Disorder."

In 1988, 290 psychiatrists . . . M. Loring and B. Powell, "Gender, Race, and DSM-III: A Study of the Objectivity of Psychiatric Diagnostic Behavior," *Journal of Health and Social Behavior* 29, no. 1 (1988): 1–22, http://dx.doi.org/10.2307/2137177.

One 2004 study showed that black men and women . . . Robert C. Schwartz and David M. Blankenship, "Racial Disparities in Psychotic Disorder Diagnosis: A Review of the Literature," *World Journal of Psychiatry* 4, no. 4 (2014): 133–40, https://www.ncbi.nlm.nih.gov/pmc/articles/PMC4274585/.

"In days of yore, most physicians" . . . Taylor, *Hippocrates Cried,* 171.

"followed dutifully in Spitzer's footsteps" . . . Scull, *Psychiatry and Its Discontents,* 284.

"godfather of medication treatment for A.D.H.D." . . . Benedict Carey, "Keith Conners, Psychologist Who Set Standard for Diagnosing A.D.H.D., Dies at 84," *New York Times,* July 13, 2017, https://nyti.ms/2viAJFe.

"The numbers make it look" . . . Carey, "Keith Conners."

"part mea culpa" . . . Frances, *Saving Normal,* xviii.

"produce a very dangerous product" . . . Frances, *Saving Normal,* xviii.

"to predict or prevent three new" . . . Frances, *Saving Normal,* 75.

childhood bipolar disorder had increased fortyfold . . . C. Moreno et al., "National Trends in the Outpatient Diagnosis and Treatment of Bipolar Disorder in Youth," *Archives of General Psychiatry* 64 (2007): 1032–39.

there had been a fifty-seven-fold increase in children's autism spectrum diagnoses . . . This number comes from comparing the 1960s/1970s numbers found in Thomas F. Boat and Joel T. Wu, eds., *Mental Disorders and Disabilities Among Low-Income Children* (Washington, DC: National Academies Press, 2015), https://www.ncbi.nlm.nih.gov/books/NBK332896/ to 2018's rates found in "Data & Statistics on Autism Spectrum Disorder," Centers for Disease Control and Prevention, https://www.cdc.gov/ncbddd/autism/data.html.

attention-deficit/hyperactivity disorder, once a rarity . . . Melissa L. Danielson et al., "Prevalence of Parent-Reported ADHD Diagnosis and Treatment Among U.S. Children and Adolescents, 2016," *Journal*

of Clinical Child & Adolescent Psychology
47, no. 2 (2018), https://www.tandfonline
.com/doi/full/10.1080/15374416.2017
.1417860.

"mislabel normal people" . . . Frances, *Saving Normal,* xviii.

"a society of pill poppers" . . . Frances, *Saving Normal,* xiv.

one in six adults . . . Thomas J. Moore and Donald R. Mattison, "Adult Utilization of Psychiatric Drugs and Differences by Sex, Age, and Race," *JAMA Internal Medicine* 177, no. 2 (2017), https://jamanetwork .com/journals/jamainternalmedicine/ fullarticle/2592697.

"an absolute scientific nightmare" . . . Scull, *Madness in Civilization,* 408.

"at best a dictionary" . . . Thomas Insel, "Post by Former NIMH Director Thomas Insel: Transforming Diagnosis," National Institute of Mental Health, April 29, 2013, https://www.nimh.nih.gov/about/directors/ thomas-insel/blog/2013/transforming -diagnosis.shtml.

I had tested this out myself . . . The SCID interview part of this chapter is from my interview with Michael First in his office on April 20, 2016.

recently that of the murder . . . James McKinley Jr., "Patz Trial Jury, in Blow to

Defense, Is Told Suspect Was a Longtime Cocaine Addict," *New York Times,* March 10, 2015, https://www.nytimes.com/2015/03/11/nyregion/patz-trial-jury-in-blow-to-defense-is-told-suspect-was-a-longtime-cocaine-addict.html.

a BBC reality show called *How Mad Are You? . . .* "How Mad Are You? Episodes 1 and 2," Horizon, BBC, November 29, 2008, https://www.bbc.co.uk/programmes/b00fm5ql.

Part Five

The greatest obstacle . . . Quotation (often misattributed to Stephen Hawking) comes from this interview with Daniel Boorstin: Carol Krucoff, "The 6 O'clock Scholar," *Washington Post,* January 29, 1984, https://www.washingtonpost.com/archive/lifestyle/1984/01/29/the-6-oclock-scholar/eed58de4-2dcb-47d2-8947-b0817a18d8fe/?utm_term=.a9cc826ca6cd. Thank you to Quote Investigator (https://quoteinvestigator.com/2016/07/20/knowledge/) for providing the proper sourcing.

22: The Footnote

The bulk of this chapter relies on several interviews with Harry Lando conducted between 2016 and 2019. I also included parts of David Rosenhan's scrap notes titled "My Basic Assumptions: Notes upon Notes" and a draft of his pseudopatient list found in his private files.

The summary read . . . Excerpt from Harry Lando, "On Being Sane in Insane Places: A Supplemental Report," *Professional Psychology,* February 1976: 47–52.

"I was the ninth pseudopatient" . . . Lando, "On Being Sane in Insane Places," 47.

"Data from a ninth pseudopatient" . . . Rosenhan, "On Being Sane in Insane Places," 258.

taught by Dr. Thelma Hunt . . . For more on Dr. Thelma Hunt, see Nicole Brigandi, "Thelma Hunt (1903–1992)," *Feminist Psychologist* 32, no. 3 (2005), https://www.apadivisions.org/division-35/about/heritage/thelma-hunt-biography.aspx.

one of her most cited works . . . Valenstein, *Great and Desperate Cures,* 165.

measuring a patient's "self-regarding span" . . . Walter Freeman and James W. Watts, *Psychosurgery: Intelligence, Emotion and Social Behavior Following Prefrontal*

Lobotomy for Mental Disorders (Spring-
field, IL: Charles C. Thomas, 1942).

his "Bobo doll study" . . . Albert Bandura,
Dorothea Ross, and Sheila A. Ross,
"Transmission of Aggression Through
Imitation of Aggressive Models," *Journal
of Abnormal and Social Psychology* 63
(1961): 575–82, https://psychclassics
.yorku.ca/Bandura/bobo.htm#f2.

"Just why Walter changed his script" . . .
David Rosenhan, "My Basic Assumptions:
Notes upon Notes," David Rosenhan
personal files.

"He engages in finger-cracking" . . . Rosenhan, "My Basic Assumptions."

talked down from the Golden Gate
Bridge . . . A few examples: "Novato Man
Held After Jump Threat," *Daily Indepen-
dent Journal,* November 2, 1964, 8; "Daly
City Wife Plucked from Golden Gate
Span," *San Mateo Times,* March 14, 1963,
24; "Model Foils S.F. Suicide," *San Mateo
Times,* June 25, 1962, 9; and "Man
Bound, Dynamite at His Throat" *Los
Angeles Times,* June 5, 1970, 146.

"Warning! Mental Patients are Notorious
DRUG EVADERS" . . . Robert Whitaker,
Mad in America, 213.

"HE LIKES IT" . . . Rosenhan, "My Basic

Assumptions."

"Didn't your dad ever teach you" . . .
Rosenhan, "My Basic Assumptions."

"I will miss it" . . . Rosenhan, "My Basic
Assumptions."

23: "It's All in Your Mind"

This chapter was based on an in-person
interview with Harry Lando in November
2016.

His hospital facilities, he revealed, were
"excellent" . . . Lando, "On Being Sane in
Insane Places," 47.

'He was admitted and diagnosed' . . .
Rosenhan, "Pseudopatient Description,"
typewritten notes, private files.

found an early draft of "On Being Sane in
Insane Places" . . . David Rosenhan, letter
to Walter Mischel, November 1971; "On
Being Sane in Insane Places," Second
Draft, David Rosenhan private files.

3.9 to 25.1 minutes . . . Rosenhan, "On Being Sane in Insane Places," 396.

"Another pseudopatient attempted a romance" . . . Rosenhan, "On Being Sane in
Insane Places," 396.

The forty-five-year-old recording opens . . .
George Bower, *It's All in Your Mind,*
WGUC-FM, December 14, 1972, NPR,

Special Collections, and university archives at the University of Maryland.

24: Shadow Mental Health Care System

She said that during this hospitalization . . . Elizabeth Lando King, phone interview, January 19, 2017.

The Zuckerberg San Francisco General Hospital . . . Thank you to the *San Francisco Gate*'s reporting for insight into what life is like at Zuckerberg San Francisco General Hospital, specifically this article: Mike Weiss, "Life and Death at San Francisco's Hospital of Last Resort," *San Francisco Gate,* December 11, 2006, https://www.sfgate.com/health/article/ GENERAL-LIFE-AND-DEATH-AT -SAN-FRANCISCO-S-2483930.php# photo-2639598.

a woman who bit off her own finger . . . Weiss, "Life and Death at San Francisco's Hospital of Last Resort."

"This is the sad part of this work" . . . Weiss, "Life and Death at San Francisco's Hospital of Last Resort."

"state of emergency" . . . . "SF General Hospital Nurses Claim Psychiatric Unit State of Emergency," KTVU, April 28, 2016, http://www.ktvu.com/news/sf

-general-hospital-nurses-claim-psychiatric
-unit-state-of-emergency.

"You've got your chow" . . . Heather Knight, "Ex-ER Psychiatrist: More Inpatient Treatment Needed in SF," *San Francisco Chronicle,* October 9, 2018, https://www .sfchronicle.com/bayarea/heatherknight/ article/Ex-ER-psychiatrist-More-inpatient -treatment-13291361.php.

"the beds that never say no" . . . Mark Gale, email to Susannah Cahalan, May 27, 2019.

"These are the choices we are making" . . . Mark Gale, phone interview, August 5, 2017.

The US is a minimum of ninety-five thousand beds . . . DJ Jaffe, *Insane Consequences: How the Mental Health Industry Fails the Mentally Ill* (Amherst, NY: Prometheus Books, 2017), 78.

It's now harder to get a bed . . . Jaffe, *Insane Consequences,* 22.

Sixty-five percent of the non-urban counties . . . C. Holly A. Andrilla, Davis G. Patterson, Lisa A. Garberson, Cynthia Coulthard, and Eric H. Larson, "Geographic Variation in the Supply of Selected Behavioral Health Providers," *American Journal of Preventive Medicine* 54, no. 6

(2018): 199–207, https://www.ajpmonline
.org/article/S0749-3797(18)30005-9/
fulltext.

national shortage of over fifteen thousand . . . Stacy Weiner, "Addressing the
Escalating Psychiatrist Shortage," *AAMC
News* (Association of American Medical
Colleges), February 13, 2018, https://news
.aamc.org/patient-care/article/addressing
-escalating-psychiatrist-shortage.

"One or more nurses would take" . . . Nathaniel Morris, "This Secret Experiment
Tricked Psychiatrists into Diagnosing
People as Having Schizophrenia," *Washington Post,* January 1, 2018.

"so disorganized that she would just
stand" . . . This psychologist prefers to
remain anonymous.

"when being assessed" . . . This nurse
prefers to remain anonymous.

"It shows just how quaint the study is" . . .
Joel Braslow, phone interview, March 11,
2015.

"It's on the other end of the spectrum" . . .
Thomas Insel, in-person interview, April
1, 2015.

A 2015 study published in *Psychiatric Services* . . . Monica Malowney, Sarah Keltz,
Daniel Fischer, and Wesley Boyd, "Avail-
ability of Outpatient Care from Psychia-

trists . . . A Simulated-Patient Study in Three Cities," *Psychiatric Services* 66, no. 1 (January 2015).

"People with schizophrenia in the United States" . . . E. Fuller Torrey, "Second Chance Lecture" at the Schizophrenia International Research Society Conference, April 1, 2016.

5 percent of people in jails . . . Torrey, *American Psychosis,* 98.

Nearly 40 percent of prisoners . . . "Indicators of Mental Health Problems Reported by Prisoners and Jail Inmates, 2011–2012," *Bureau of Justice Statistics* (2017), https://www.bjs.gov/content/pub/pdf/imhprpji1112_sum.pdf.

Women, the fastest growing segment . . . "Indicators of Mental Health Problems," Bureau of Justice.

"are more likely to suffer disparities" . . . Lorna Collier, "Incarceration Nation," *American Psychological Association* 45, no. 9 (2014): 56, https://www.apa.org/monitor/2014/10/incarceration.

ten times more seriously mentally ill people . . . "Serious Mental Illness (SMI) Prevalence in Jails and Prisons," Treatment Advocacy Center Office of Research and Public Affairs, September 2016, https://www.treatmentadvocacycenter.org/

storage/documents/ backgrounders/smi-in-jails-and-prisons.pdf.

The largest concentrations of the seriously mentally ill . . . "Serious Mental Illness," Treatment Advocacy Center; and Gale Holland, "L.A. County Agrees to New Policies to End the Jail-to-Skid Row Cycle for Mentally Ill People," *LA Times,* December 7, 2018, https://www.latimes.com/local/lanow/la-me-ln-skid-row-jail-2018 1207-story.html.

"Many of the persons with serious mental illness" . . . Richard Lamb, in-person interview, October 29, 2015.

This is the current state . . . Some have argued that the clear-cut connection between deinstitutionalization and transinstitutionalization is oversimplified. For a more nuanced perspective on the history of incarceration, see Michelle Alexander, *The New Jim Crow: Mass Incarceration in the Age of Colorblindness* (New York: New Press, 2012); Bryan Stevenson, *Just Mercy: A Story of Justice and Redemption* (New York: Spiegel & Grau, 2014); and John Pfaff, *Locked In: The True Causes of Mass Incarceration — And How to Achieve Real Reform* (New York: Basic Books, 2017).

"A crisis unimaginable" . . . Powers, *No One Cares About Crazy People,* 203.

"one of the greatest social debacles" . . . Shorter, *A History of Psychiatry,* 277.

"a cruel embarrassment" . . . "Denying the Mentally Ill" (editorial), *New York Times,* June 5, 1981, https://www.nytimes.com/1981/06/05/opinion/denying-the-mentally-ill.html.

"Behind the bars of prisons and jails" . . . Dominic Sisti, "Psychiatric Institutions Are a Necessity," *New York Times,* May 9, 2016, https://www.nytimes.com/roomfordebate/2016/05/09/getting-the-mentally-ill-out-of-jail-and-off-the-streets/ psychiatric-institutions-are-a-necessity.

the average stay for a mentally ill prisoner . . . E. T. Torrey, M. T. Zdanowicz, A. D. Kennard, "The Treatment of Persons with Mental Illness in Prisons and Jails: A State Survey," *Treatment Advocacy Center,* April 8, 2014, https://www.treatmentadvocacycenter.org/storage/documents/backgrounders/how%20many%20individuals%20with%20serious%20mental%20illness%20are%20in%20jails%20and%20pris ons%20final.pdf.

The ACLU filed a lawsuit . . . *J.H. v. Miller.*

languished in jail for 1,017 days . . . "Lawsuit Alleges Many Defendants with Mental Illness Jailed for Well Over a Year Awaiting

Mental Health Treatment," *ACLU Pennsylvania,* October 22, 2015, https://www
.aclupa.org/news/2015/10/22/lawsuit
-alleges-many-defendants-mental-illness
-jailed-well-o.

The lawsuit's lead plaintiff is "J.H." . . .
"J.H. v. Miller (Formerly J.H. v. Dallas),"
ACLU Pennsylvania, October 22, 2015,
https://www.aclupa.org/our-work/legal/
legaldocket/jh-v-dallas.

"failed to produce constitutionally" . . .
"ACLU-PA Goes Back to Court on Behalf
of People Who Are Too Ill to Stand Trial,"
ACLU Pennsylvania, March 19, 2019,
https://www.aclupa.org/news/2019/03/19/
aclu-pa-goes-back-court-behalf-people
-who-are-too-ill-stand.

"often nude, are covered in filth" . . . Eric
Balaban, "Time Has Come to Save Mentally Ill Inmates from Solitary Confinement" (editorial), *Arizona Capital Times,*
February 27, 2018, https://azcapitoltimes
.com/news/2018/02/27/time-has-come-to
-save-mentally-ill-inmates-from-solitary
-confinement.

In California, "Inmate Patient X" . . . Hannah Fry, "Inmate Rips Out Her Own Eye
and Eats It: Report Slams Mental Healthcare in California Prisons," *Los Angeles
Times,* November 5, 2018, https://www

.latimes.com/local/lanow/la-me-ln-prison
-report-20181105-story.html.

In Florida, Darren Rainey . . . Roth, *Insane,*
135.

In Mississippi, "a real 19th century hell
hole" . . . Craig Haney, "Madness and
Penal Confinement: Observations on
Mental Illness and Prison Pain," Draft,
provided to me by Craig Haney.

a man named Michael Tyree screamed
out . . . Tracey Kaplan, "Guard Trial: Fellow Inmate Testifies Michael Tyree Was
'Screaming for His Life,' " *Mercury News,*
March 23, 2017, https://www.mercury
news.com/2017/03/23/jail-trial-testimony
-over-inmate-death-probes-delay-sum
moning-help-for-michael-tyree.

"I have seen them" . . . J. E. D. Esquirol,
"Des établissemens des aliénés en France,
et des moyens d'améliorer le sort de ces
infortunés: Mémoire présenté à Son Excellence le ministre de l'intérieur, en sep-
tembre 1818," reprinted in Mark S. Mi-
cale and Roy Porter, eds., *Discovering the
History of Psychiatry* (Oxford: Oxford
University Press, 1994), 235.

"It's true that the *hospitals*" . . . Roth,
Insane, 2.

"Prisoners are under a tremendous amount
of stress" . . . Craig Haney, in-person

interview, February 17, 2017.

"How do you know when a patient is lying?" . . . Jimmy Jenkins, "Whistleblower: Patients with Mental Illness Suffering in Arizona" (radio program), KJZZ, June 1, 2018, https://kjzz.org/content/644690/whistleblower-patients-mental-illness-suffering-arizona-prisons.

"I mean people who have documented histories" . . . David Fathi, phone interview, April 7, 2015.

"What's the secondary gain?" . . . Craig Haney, in-person interview, February 17, 2017.

Dr. Torrey, the psychiatrist who warned . . . Thank you to Dr. Torrey and to DJ Jaffe for taking time to speak to me about these issues. For more on Dr. Torrey's perspective, see his large body of work, including some of his books cited here: *American Psychosis, Surviving Schizophrenia, The Insanity Offense,* and *Out of the Shadows.* For more from DJ Jaffe, see https://mentalillnesspolicy.org/ and his book *Insane Consequences.* For a great summary of DJ Jaffe's solutions to these many issues in New York City see DJ Jaffe and Stephen Eide, "How to Fix New York's Mental Health Crisis Without Spending More Money," *New York Post,* May 11,

2019, https://nypost.com/2019/05/11/how
-to-fix-new-yorks-mental-health-crisis
-without-spending-more-money/.

adding more beds across the board . . .
Doris A. Fuller, Elizabeth Sinclair,
H. Richard Lamb, James D. Cayce, and
John Snook, "Emptying the 'New
Asylums': A Beds Capacity Model to
Reduce Mental Illness Behind Bars,"
Treatment Advocacy Center, January 2017,
https://www.treatmentadvocacycenter.org/
storage/documents/emptying-new-asylums
.pdf.

"human trigger warning" . . . DJ Jaffe,
"Insane Consequences: How the Mental
Health Industry Fails the Mentally Ill,"
TEDx at the National Council of Behavioral Health, April 25, 2018, https://
mentalillnesspolicy.org/tedtalk-and-op
-eds/.

more mental health courts . . . Jaffe, *Insane
Consequences,* 233–34.

crisis intervention teams . . . Jaffe, *Insane
Consequences,* 232–33.

using legal force to get people to take their
meds . . . Jaffe, *Insane Consequences,*
234–35.

civil commitment reforms . . . "Improving
Civil Commitment Laws and Standards,"
Treatment Advocacy Center, https://www

.treatmentadvocacycenter.org/fixing-the
-system/improving-laws-and-standards.

a small subset of people, who are typically untreated . . . E. Fuller Torrey, "Stigma and Violence: Isn't It Time to Connect the Dots?" *Schizophrenia Bulletin* 37, no. 5 (2011): 892–96, https://www.ncbi.nlm.nih .gov/pmc/articles/PMC3160234/.

"Being psychotic is not an exercise" . . . DJ Jaffe is quoted in Carrie Arnold, "How Do You Treat Someone Who Doesn't Accept They're Ill?" BBC, August 7, 2018, http://www.bbc.com/future/story/2018 0806-how-do-you-treat-someone-who -doesnt-accept-theyre-ill.

Sheriff Tom Dart of Chicago's Cook County jail . . . Lesley Stahl, "Half of the Inmates Shouldn't Be Here, Says Cook County Sheriff," *60 Minutes,* May 21, 2017, https://www.cbsnews.com/news/cook -county-jail-sheriff-tom-dart-on-60 -minutes/.

"If I told you that was the case for cancer" . . . Thomas Insel, in-person interview, April 1, 2015.

25: The Hammer

social constructionist . . . Girishwar Misra and Anand Prakash, "Kenneth J. Gergen and Social Constructionism," *Psychological Studies* 57, no. 2 (2012): 121–25, https://link.springer.com/article/10.1007/s12646-012-0151-0.

"To meet [Rosenhan] and talk with him" . . . Kenneth Gergen, phone interview, January 17, 2016.

We discussed her eclectic work . . . Nancy Horn, phone interviews, November 3, 2015; February 25, 2015; March 13, 2015; and in-person, April 14, 2015.

protect "the rights and welfare" . . . "Institutional Review Boards Frequently Asked Questions," U.S. Food & Drug Administration (1998), https://www.fda.gov/regulatory-information/search-fda-guidance-documents/institutional-review-boards-frequently-asked-questions.

Chestnut Lodge was a famous private psychiatric hospital . . . The history of Chestnut Lodge was culled from a variety of sources, among them Ann-Louise S. Silver, "Chestnut Lodge, Then and Now," *Contemporary Psychoanalysis* 33, no. 2 (1997): 227–49; Neal Fitzsimmons, "Woodlawn Hotel — Chestnut Lodge Sanitarium, the Bullard Dynasty," *Montgomery County Historical Society* 17, no. 4 (1974): 2–11; and interviews with former staff, including a phone interview with

Cindy Sargent on October 6, 2015, and an in-person interview with Pamela Shell on June 15, 2015.

Dr. Ray Osheroff, a depressed forty-one-year-old . . . The history of Dr. Ray Osheroff came from Mark Moran, "Recalling Chestnut Lodge: Seeking the Human Behind the Psychosis," *Psychiatric News,* April 25, 2014, https://psychnews.psy chiatryonline.org/doi/10.1176/appi.pn .2014.5a17; Sandra G. Boodman, " 'A Horrible Place, a Wonderful Place,' " *Washington Post,* October 8, 1989, https:// www.washingtonpost.com; and Sharon Packer, "A Belated Obituary: Raphael J. Osheroff, MD," *Psychiatric Times,* June 28, 2013, http://www.psychiatrictimes .com/blog/belated-obituary-raphael-j -osheroff-md.

"psychiatry was a house divided" . . . Packer, "A Belated Obituary."

Then, on July 13, 2009 . . . Asha Beh, "Historic Rockville Asylum Destroyed in Two-Alarm Fire," NBC Washington, July 13, 2009, https://www.nbcwashington .com/news/local/Historic-Rockville -Asylum-Destroyed-in-Two-Alarm-Fire .html.

"This is a summertime photo" . . . The interviewee wishes to remain anonymous.

641

Laura did use the opportunity . . . Rosenhan, *Odyssey into Lunacy,* chapter 6, 13.

"I didn't take part in this study" . . . Judith Godwin, email to Susannah Cahalan, February 9, 2016.

Grace Hartigan, who was born in Newark . . . Grace Hartigan's history was compiled from a variety of sources, including Cathy Curtis, *Restless Ambition* (Oxford: Oxford University Press, 2015); William Grimes, "Grace Hartigan, 86, Abstract Painter, Dies," *New York Times,* November 18, 2008, https://www.nytimes .com/2008/11/18/arts/design/18hartigan .html; and Michael McNay, "Grace Hartigan," *The Guardian,* November 23, 2008, https://www.theguardian.com/artanddes ign/2008/nov/24/1. Also helpful were phone interviews with Cathy Curtis (February 8, 2016); Daniel Belasco (February 11, 2015); and Hart Perry (February 12, 2016).

"It's not Grace" . . . Rex Stevens, phone interview, February 14, 2016.

Excerpt of David Rosenhan outline for his unpublished book, from his private files.

a series of letters written by a woman . . . Letters between Mary Peterson and David Rosenhan can be found in the David L. Rosenhan Papers.

self-published book of adoring short stories . . . Mary Pledge Peterson, *Life Is So Daily in Cincinnati* (Cincinnati: Cincinnati Book Publishers, 2012).

"An angel on wheels" . . . Phil Nuxhall, "An Angel on Wheels," *Positive 365,* 2012, http://www.positive365.com/Positive -Magazine/Positive-2012/An-Angel-on -Wheels.

"gray-haired" and "grandmotherly" . . . Rosenhan, *Odyssey into Lunacy,* chapter 3, 16.

I contacted Mary's surviving sister and childhood best friend . . . Betty Pledge Maxey, phone interview, January 13, 2016; and Connie Selvey, phone interview, January 26, 2016.

"There's no way that Mary was a pseudopatient" . . . Florence Keller, phone interview, March 26, 2016.

"the founding father of positive psychology" . . . "The 5 Founding Fathers of Positive Psychology," Positive Psychology Program, February 8, 2019, https:// positivepsychologyprogram.com/founding -fathers.

His biography matched up . . . For more on Seligman, see his memoir, *The Hope Circuit: A Psychologist's Journey from Helplessness to Optimism* (New York: Public

Affairs, 2018).

he did go undercover at Norristown State Hospital . . . Medical records and letters recording Rosenhan's and Seligman's stay at Norristown can be found in the David L. Rosenhan Papers.

thirty-eight and forty-eight . . . Rosenhan lists various ages for Carl in different locations, such as his unpublished book and his pseudopatient list.

but he had died in 1992 . . . Bruce Lambert, "Perry London, 61, Psychologist; Noted for His Studies of Altruism," *New York Times,* June 22, 1992, https://www.nytimes .com/1992/06/22/nyregion/perry-london -61-psychologist-noted-for-his-studies-of -altruism.html.

His daughter Miv, a psychotherapist . . . Miv London, phone interview, February 8, 2016.

"Everyone loved David" . . . Vivian London, Skype interviews, February 8, 2016, and March 3, 2016.

"It has become obvious" . . . Vivian London, email to Susannah Cahalan, February 8, 2016.

wrote a letter of recommendation for Leibovitz . . . David Rosenhan, letter to David Hapgood, November 4, 1970, David L. Rosenhan Papers.

a glowing *New York Times* obituary . . . "Dr. Maury Leibovitz, Art Dealer and Clinical Psychologist, 75," *New York Times,* June 5, 1992, https://www.nytimes.com/1992/06/05/arts/dr-maury-leibovitz-art-dealer-and-a-clinical-psychologist-75.html.

The next day a man's Southern California drawl . . . Josh Leibovitz, phone interview, February 10, 2016.

"No one with the name or initials" . . . text message to Susannah Cahalan, February 13, 2016.

"I spoke with mother" . . . Josh Leibovitz, email to Susannah Cahalan, March 2, 2016.

"The upper portion of the painting" . . . Rosenhan, *Odyssey into Lunacy,* chapter 6, 16–17.

"The bottom half of the painting [is] much less intense" . . . Rosenhan, *Odyssey into Lunacy,* chapter 6, 18–19.

26: An Epidemic

I wrote a commentary . . . Susannah Cahalan, "In Search of Insane Places" (correspondence), *Lancet Psychiatry* 4, no. 5 (2017), http://dx.doi.org/10.1016/S2215-0366(17)30138-4.

"Well, he did often use some" . . . Carole

Westmoreland, phone interview, December 5, 2016.

"minimal self-references and convoluted phrases" . . . Sarah Griffiths, "The Language of Lying," *Daily Mail,* November 5, 2014, http://www.dailymail.co.uk/sciencetech/article-2821767/The-language-LYING-Expert-reveals-tiny-clues-way-people-talk-reveal-withholding-truth.html.

he said that it was impossible to suss . . . Jamie Pennebaker, phone interview, May 2017.

"I continue to wonder" . . . Florence Keller, email to Susannah Cahalan, February 15, 2017.

His publisher, Doubleday, sued him . . . *Doubleday & Company, Inc. v. David L. Rosenhan.*

publishing a paper on the effects of success . . . Isen, Horn, and Rosenhan, "Effects of Success and Failure on Children's Generosity."

mood and self-gratification . . . Underwood, Moore, and Rosenhan, "Affect and Self-Gratification."

joys of helping . . . David L. Rosenhan, Peter Salovey, and Kenneth Hargis, "The Joys of Helping: Focus of Attention Mediates the Impact of Positive Affect on Altruism," *Journal of Personality and Social Psychology* 40, no. 5 (1981): 899–905.

moral character . . . David L. Rosenhan, "Moral Character," *Stanford Law Review* 27, no. 3 (1975): 925–35.

pseudoempiricism . . . David L. Rosenhan, "Pseudoempiricism: Who Owns the Right to Scientific Reality?" *Psychological Inquiry* 2, no. 4 (1991): 361–63.

study of nightmares experienced after an earthquake . . . James M. Wood, Richard R. Bootzin, David Rosenhan, Susan Nolen-Hoeksema, and Forest Jourden, "Effects of 1989 San Francisco Earthquake on Frequency and Content of Nightmares," *Journal of Abnormal Psychology* 101, no. 2 (1992): 219–24.

"David became sort of less" . . . Michael Wald, phone interview, February 16, 2016.

one paper on how notetaking aids jurors' recall . . . David L. Rosenhan, Sara L. Eisner, and Robert J. Robinson, "Notetaking Aids Juror Recall," *Law and Human Behavior* 18, no. 1 (1994): 53–61.

on their ability (or, rather, inability) to disregard facts . . . William C. Thomson, Geoffrey T. Fong, and David L. Rosenhan, "Inadmissible Evidence and Jury Verdicts," *Journal of Personality and Social*

Psychology 40, no. 3 (1981): 453–63.

a shocking percentage of Stanford students . . . David Rosenhan, "Intense Religiosity," Comment Draft, unpublished, accessed from private files.

"Whenever you'd try to find him" . . . The former graduate student prefers to remain anonymous.

"I was suspicious of him" . . . Eleanor Maccoby, in-person interview, February 22, 2017.

"I never really connected with Rosenhan" . . . Walter Mischel to Lee Ross, email, forwarded to Susannah Cahalan, February 15, 2017.

"He could make you feel" . . . This person prefers to remain anonymous.

"absolutely not possible" . . . Nancy Horn, phone interview, May 13, 2019.

"My dad was a storyteller" . . . Jack Rosenhan, in-person interview, February 20, 2017.

"I don't know" . . . Bill and Maryon Underwood, phone interview, July 8, 2016.

"I never thought of him as a BS artist" . . . Harry Lando, in-person interview, November 19, 2016.

social psychologist Diederik Stapel . . . For a great summation of Stapel's fraud, see Yudhijit Bhattacharjee, "The Mind of a

Con Man," *New York Times,* April 26, 2013, https://www.nytimes.com/2013/04/28/magazine/diederik-stapels-audacious-academic-fraud.html; and Martin Enserink, "Dutch University Sacks Social Psychologist over Faked Data," *Science News,* September 7, 2011, https://www.sciencemag.org/news/2011/09/dutch-university-sacks-social-psychologist-over-faked-data.

published in *Science* about a correlation . . . D. A. Stapel and S. Lindenberg, "Coping with Chaos: How Disordered Contexts Promote Stereotyping and Discrimination," *Science* 332 (2011): 251–53.

"perhaps the biggest con man" . . . Bhattacharjee, "The Mind of a Con Man."

this level of con *could* happen . . . For a great rundown of how this level of con happens in academia, read Richard Harris, *Rigor Mortis: How Sloppy Science Creates Worthless Cures, Crushes Hope, and Wastes Billions* (New York: Basic Books, 2017).

midst of a "replication crisis" . . . Ed Yong, "Psychology's Replication Crisis Is Running Out of Excuses," *The Atlantic,* November 19, 2018, https://www.theatlantic.com/science/archive/2018/11/psychologys-replication-crisis-real/576223/.

"power posing" . . . Susan Dominus, "When the Revolution Came for Amy Cuddy," *New York Times,* October 18, 2017, https://www.nytimes.com/2017/10/18/magazine/when-the-revolution-came-for-amy-cuddy.html.

"the facial feedback hypothesis" . . . Stephanie Pappas, "Turns Out, Faking a Smile Might Not Make You Happier After All," *LiveScience,* November 3, 2016, https://www.livescience.com/56740-facial-feedback-hypothesis-fails-in-replication-attempt.html.

"ego depletion" . . . Daniel Engber, "Everything Is Crumbling," *Slate,* March 6, 2016, http://www.slate.com/articles/health_and_science/cover_story/2016/03/ego_depletion_an_influential_theory_in_psychology_may_have_just_been_debunked.html.

started the "Reproducibility Project" . . . "Estimating the Reproducibility of Psychological Science," *Science* 349, no. 6251 (August 28, 2015): 943–53, http://science.sciencemag.org/content/349/6251/aac4716/tab-pdf.

A replication of the study . . . Tyler W. Watts, Greg J. Duncan, and Haonan Quan, "Revisiting the Marshmallow Test: A Conceptual Replication Investigating Links Between Early Delay of Gratification and Later Outcomes," *Psychological Science* 29, no. 7 (2018), https://doi.org/10.1177/0956797618761661.

Yet the marshmallow test and its follow-ups . . . Brian Resnick, "The 'Marshmallow Test' Said Patience Was a Key to Success. A New Replication Tell Us S'More," *Vox,* June 8, 2018, https://www.vox.com/science-and-health/2018/6/6/17413000/marshmallow-test-replication-mischel-psychology.

Stanley Milgram and his shock tests . . . Perry, *Behind the Shock Machine.*

including a 2017 paper out of Poland . . . Dariusz Dolinski, Tomasz Grzyb, Michal Folwarczny, "Would You Deliver an Electric Shock in 2015? Obedience in Experimental Paradigm Developed by Stanley Milgram in the Fifty Years Following the Original Study," *Social Psychological and Personality Science* 8, no. 8 (2017): 927–33, https://journals.sagepub.com/doi/10.1177/1948550617693060.

Among the hardest hit . . . Thank you to Philip Zimbardo for taking the time to speak with me on Skype, October 2, 2015.

recruited students from a newspaper ad . . . Haney, Banks, and Zimbardo, "Interpersonal Dynamics in a Simulated Prison."

"I was shocked. But not surprised" . . . Claudia Dreifus, "Finding Hope in Knowing the Universal Capacity for Evil," *New York Times,* April 3, 2007, https://www.nytimes.com/2007/04/03/science/03conv.html.

"not reformable" . . . Ben Blum, "The Lifespan of a Lie," *Medium,* June 7, 2018, https://medium.com/s/trustissues/the-lifespan-of-a-lie-d869212b1f62.

"It was just a job" . . . Blum, "The Lifespan of a Lie."

"We must stop celebrating this work" . . . Brian Resnick, "The Stanford Prison Study Was Massively Influential. We Just Found Out It Was a Fraud," *Vox,* June 13, 2018, https://www.vox.com/2018/6/13/17449118/stanford-prison-experiment-fraud-psychology-replication.

"prime example of a study that fits our biases" . . . Peter Gray, phone interview, December 28, 2016.

Caroline Barwood and colleague Bruce Murdoch . . . "Ex-UQ Academic Found Guilty of Fraud," 9News.com, October 24, 2016, https://www.9news.com.au/national/2016/10/24/17/05/ex-uq-academic-found-guilty-of-fraud.

Korean stem-cell researcher Hwang Woo Suk . . . Choe Sang-Hun, "Disgraced

Cloning Expert Convicted in South Korea," *New York Times,* October 26, 2009, https://www.nytimes.com/2009/10/27/world/asia/27clone.html.

There's Elizabeth Holmes . . . For a rollercoaster ride of a story on the Theranos scandal, see John Carreyrou, *Bad Blood: Secrets and Lies in a Silicon Valley Startup* (New York: Knopf, 2018).

"Much of the scientific literature" . . . Richard Horton, "Offline: What Is Medicine's 5 Sigma?" *Lancet* 385 (2015), https://www.thelancet.com/journals/lancet/article/PIIS0140-6736(15)60696-1/fulltext.

One of the leaders of the push to uncover academic fraud . . . John P. A. Ioannidis, "Why Most Published Research Findings Are False," *PLOS Medicine* 2, no. 8 (2005), https://journals.plos.org/plosmedicine/article?id=10.1371/journal.pmed.0020124.

He's found that out of thousands of early papers . . . John P. A. Ioannidis, Robert Tarone, and Joseph K. McLaughlin, "The False-Positive to False-Negative Epidemiological Studies," *Epidemiology* 22, no. 4 (2011): 450–56, https://www.gwern.net/docs/statistics/decision/2011-ioannidis.pdf.

followed forty-nine studies . . . Ben Goldacre, "Studies of Studies Show That We Get Things Wrong," *The Guardian,* July 15, 2011, https://www.theguardian.com/commentisfree/2011/jul/15/bad-science-studies-show-we-get-things-wrong.

"flatly contradicted" . . . Goldacre, "Studies of Studies."

Brian Wansink resigned . . . Eli Rosenberg and Herman Wong, "This Ivy League Food Scientist Was a Media Darling. He Just Submitted His Resignation, School Says," *Washington Post,* September 20, 2018, https://www.washingtonpost.com/health/2018/09/20/this-ivy-league-food-scientist-was-media-darling-now-his-studies-are-being-retracted/?utm_term=.4457b7c5cb0b.

"academic misconduct in his research" . . . Michael I. Kotlikoff, "Statement of Cornell University Provost Michael I. Kotlikoff," Cornell University, September 20, 2018, https://statements.cornell.edu/2018/20180920-statement-provost-michael-kotlikoff.cfm.

"falsified and/or fabricated data" . . . Gina Kolata, "Harvard Calls for Retraction of Dozens of Studies by Noted Cardiac Researcher," *New York Times,* October 15, 2018, https://www.nytimes.com/2018/10/

15/health/piero-anversa-fraud-retractions
.html.

the fraudulent Wakefield study . . . The original study, since retracted, is A. J. Wakefield, S. H. Murch, A. Anthony, J. Linnell, D. M. Casson, M. Malik, et al., "Ileal Lymphoid Nodular Hyperplasia, Non-specific Colitis, and Pervasive Developmental Disorder in Children," *Lancet* 351 (1998): 637–41. The definitive paper that exposed the study's fraud is Editors, "Wakefield's Article Linking MMR Vaccine and Autism Was Fraudulent," *BMJ* (2011), https://www.bmj.com/content/342/bmj.c7452.full.print#ref-2.

"palliative, none are even proposed as cures" . . . T. R. Insel and E. M. Scolnick, "Cure Therapeutics and Strategic Prevention: Raising the Bar for Mental Health Research," *Molecular Psychiatry* 11 (2006): 13.

Second-generation drugs . . . An NIMH study, called the Clinical Antipsychotic Trials of Intervention Effectiveness (CATIE), compared older drugs with atypical antipsychotics and found that "the newer drugs were no more effective or better tolerated than the older drugs" with the exception of one, Clozapine. "Questions and Answers About the NIMH

Clinical Antipsychotic Trials of Intervention Effectiveness Study (CATIE) — Hase 2 Results," National Institute of Mental Health, https://www.nimh.nih.gov/funding/clinical-research/practical/catie/phase2results.shtml.

"the single biggest target" . . . Duff Wilson, "Side Effects May Include Lawsuits," *New York Times,* October 2, 2010, https://www.nytimes.com/2010/10/03/business/03psych.html.

Johnson & Johnson, for example . . . Katie Thomas, "J&J to Pay $2.2 Billion in Risperdal Settlement," *New York Times,* November 4, 2013, https://www.nytimes.com/2013/11/05/business/johnson-johnson-to-settle-risperdal-improper-marketing-case.html.

"For the past twenty-five years" . . . Robert Whitaker, *Anatomy of an Epidemic: Magic Bullets, Psychiatric Drugs, and the Astonishing Rise of Mental Illness in America* (New York: Crown, 2010), 358.

"They just need to take their drugs" . . . Psychiatrist, in-person interview.

"your life is taken away from you" . . . This person prefers to remain anonymous.

I see that these drugs help many people . . . For a remarkable story about how the right antipsychotic medication (in this

656

case Clozapine) helped turn a life around, see Bethany Yeiser's *Mind Estranged: My Journey from Schizophrenia and Homelessness to Recovery* (2014).

a worldwide shortage of mental health care workers . . . Kitty Farooq et al., "Why Medical Students Choose Psychiatry — A 20 Country Cross-Sectional Survey," *BMC Medical Education* 14, no. 12 (2014), https://bmcmededuc.biomedcentral.com/articles/10.1186/1472-6920-14-12.

only 3 percent of Americans . . . M. M. Weissman, H. Verdeli, S. E. Bledsoe, K. Betts, H. Fitterling, and P. Wickramaratne, "National Survey of Psychotherapy Training in Psychiatry, Psychology, and Social Work," *Archives of General Psychiatry* 63, no. 8 (2006): 925–34, https://www.ncbi.nlm.nih.gov/pubmed/16894069.

"Before we get to that" . . . Allen Frances, phone interview, January 4, 2016.

27: Moons of Jupiter

Taunted by death . . . Rita Charon and Peter Wyer, "The Art of Medicine," *Lancet* 371(2008): 296–97, https://www.thelancet.com/pdfs/journals/lancet/PIIS0140-6736(08)60156-7.pdf.

"I think we should be honest about" . . . Belinda Lennox, phone interview, December 29, 2016.

Dutch psychiatrist Jim van Os, who wrote . . . S. Guloksuz and J. van Os, "The Slow Death of the Concept of Schizophrenia and the Painful Birth of the Psychosis Spectrum," *Psychology Medicine* 48, no. 2 (2018): 229–44, https://www.ncbi.nlm.nih.gov/pubmed/28689498.

"not more than ten diagnoses" . . . Jim van Os, phone interview, August 3, 2017.

The research community has reached . . . In Japan, psychiatrists replaced the term *Seishin Bunretsu Byo* (mind-split disease) with *Togo Shitcho Sho* (integration disorder) in 2002. There's evidence that this change in the nomenclature has opened up better communication channels between doctors and patients: Before the change, only 7 percent of psychiatrists always shared diagnosis with patients; within seven months, 78 percent of psychiatrists did.

"Is schizophrenia disappearing?" . . . Per Bergsholm, "Is Schizophrenia Disappearing?" *BMC Psychiatry* 16 (2016), https://bmcpsychiatry.biomedcentral.com/articles/10.1186/s12888-016-1101-5.

"Should the label schizophrenia be abandoned?" . . . A. Lasalvia, E. Penta, N. Sartorius, and S. Henderson, "Should the Label Schizophrenia Be Abandoned?" *Schizophrenia Research* 162, nos. 1–3 (2015): 276–84, https://www.ncbi.nlm.nih.gov/pubmed/25649288.

During his tenure as the director . . . My understanding of the *RDoC* came from a variety of sources, but was mainly compiled from an in-person interview on June 15, 2015, and "*Research Domain Criteria (RDoC)*," National Institute of Mental Health, https://www.nimh.nih.gov/research/research-funded-by-nimh/rdoc/index.shtml.

half of NIMH-funded studies . . . Sarah Deweerdt, "US Institute Maintains Support for Diagnoses Based on Biology," *Spectrum,* May 9, 2018. For more on RDoC criteria, see https://www.psychiatrictimes.com/nimh-research-domain-criteria-rdoc-new-concepts-mental-disorders.

from 10 to 30 percent . . . Frederick J. Frese, Edward L. Knight, and Elyn Saks, "Recovery from Schizophrenia: With Views of Psychiatrists, Psychologists, and Others Diagnosed with This Disorder," *Schizophrenia Bulletin* 35, no. 2 (2009): 370–80, https://www.ncbi.nlm.nih.gov/

pmc/articles/PMC2659312/.

but hundreds . . . Linda Geddes, "Huge Brain Study Uncovers 'Buried' Genetic Networks Linked to Mental Illness," *Nature News,* December 13, 2018, https://www.nature.com/articles/d41586-018-07750-x.

a "genetic overlap" in psychiatric disorders . . . The Brainstorm Consortium, "Analysis of Shared Heritability in Common Disorders of the Brain," *Science* 360, no. 6395 (2018), https://www.ncbi.nlm.nih.gov/pmc/articles/PMC6097237/; and Alastair G. Cardno and Michael J. Owen, "Genetic Relationship Between Schizophrenia, Bipolar Disorder, and Schizoaffective Disorder," *Schizophrenia Bulletin* 40, no. 3 (2014): 504–15, https://www.ncbi.nlm.nih.gov/pmc/articles/PMC3984527/.

"The tradition of drawing these sharp lines . . . Karen Zusi, "Psychiatric Disorders Share an Underlying Genetic Basis," *Science Daily,* June 21, 2018, https://www.sciencedaily.com/releases/2018/06/180621141059.htm.

spurring studies of immune-suppressing drugs . . . One such example comes out of Oxford University: Belinda R. Lennox, Emma C. Palmer-Cooper, Thomas Pollack, Jane Hainsworth, Jacqui Marks, Leslie Jacobson, "Prevalence and Clinical Characteristics of Serum Neuronal Cell Surface Antibodies in First-Episode Psychosis: A Case-Control Study," *Lancet Psychiatry* 4, no. 1 (2017): 42–48, https://www.thelancet.com/journals/lanpsy/article/PIIS2215-0366%2816%2930375-3/fulltext.

a third of people with schizophrenia . . . Moises Velasquez-Manoff, "He Got Schizophrenia. He Got Cancer. And Then He Got Cured," *New York Times,* September 29, 2018, https://www.nytimes.com/2018/09/29/opinion/sunday/schizophrenia-psychiatric-disorders-immune-system.html.

reduce mania . . . F. Dickerson et al., "Adjunctive Probiotic Microorganism to Prevent Rehospitalization in Patients with Acute Mania: A Randomized Control Trial," *Bipolar Disorders* 20, no. 7 (2018): 614–21.

the more robust symptoms of schizophrenia . . . Emily G. Severance et al., "Probiotic Normalization of *Candida albicans* in Schizophrenia: A Randomized, Placebo-Controlled Longitudinal Pilot Study," *Brain Behavior and Immunity* 62 (2017): 41–45.

people born in winter months . . . Erick

Messias, Chuan-Yu Chen, and William W. Eaton, "Epidemiology of Schizophrenia: Review of Findings and Myths," *Psychiatric Clinics of North America* 8, no. 9 (2011): 14–19, https://www.ncbi.nlm.nih.gov/pmc/articles/PMC3196325/.

are more likely to be born in the summer . . . Thank you, Dr. William Carpenter, for the heads-up about this. Erick Messias, Brian Kirkpatrick, and Evelyn Bromet, "Summer Birth and Deficit Schizophrenia: A Pooled Analysis from Six Countries," *JAMA Psychiatry* 61, no. 10 (2004): 985–99, https://jamanetwork.com/journals/jamapsychiatry/fullarticle/482066.

"What I teach my students is" . . . Steven Hyman, phone interview, February 10, 2017.

a highly touted paper in *Nature* . . . Aswin Ekar et al., "Schizophrenia Risk from Complex Variation of Complement Component 4," *Nature* 530 (2016): 177–83, https://www.nature.com/articles/nature16549.

Drop-Seq . . . Lisa Girard, "Single-Cell Analysis Hits Its Stride: Advances in Technology and Computational Analysis Enable Scale and Affordability, Paving the Way for Translational Studies," Broad

Institute, May 21, 2015, https://www
.broadinstitute.org/news/single-cell
-analysis-hits-its-stride.

optogenetics, which manipulates brain circuits . . . Stephen S. Hall, "Neuroscience's New Toolbox," *MIT Technology Review,* June 17, 2014, https://www.tech nologyreview.com/s/528226/neurosciences -new-toolbox.

CLARITY, which melts away the superstructure . . . Mo Costandi, "CLARITY Gives a Clear View of the Brain," *The Guardian,* April 10, 2013, https://www .theguardian.com/science/neurophilos ophy/2013/apr/10/clarity-gives-a-clear -view-of-the-brain.

a new technique . . . Ruixan Gao et al., "Cortical Column and Whole-Brain Imaging with Molecular Contrast and Nanoscale Resolution," *Science* 363, no. 6424 (2019), https://science.sciencemag.org/ content/363/6424/eaau8302.

They are in essence creating "minibrains" . . . Dina Fine Maron, "Getting to the Root of the Problem: Stem Cells Are Revealing New Secrets About Mental Illness," *Scientific American,* February 27, 2018, https://www.scientificamerican.com/ article/getting-to-the-root-of-the-problem -stem-cells-are-revealing-new-secrets

-about-mental-illness.

IBM's Watson team told me . . . I visited the facility and received a tour from Guillermo Cecchi and company on November 16, 2016.

"Digital phenotyping" . . . Thomas R. Insel, "Digital Phenotyping: A Global Tool for Psychiatry," *World Psychiatry* 17, no. 3 (2018): 276–78, https://www.ncbi.nlm.nih .gov/pmc/articles/PMC6127813/.

More medical students are pursuing careers . . . Mark Moran, "U.S. Seniors Matching to Psychiatry Increases for Sixth Straight Year," *Psychiatric News,* American Psychiatric Association, March 29, 2018, https://doi.org/10.1176/appi.pn.2018.4a.

the average psychiatrist's salary increased . . . Carol Peckham, "Medscape Psychiatrist Compensation Report 2018," *Medscape,* April 18, 2018, https://www .medscape.com/slideshow/2018-com pensation-psychiatrist-6009671#8.

"We have never seen demand" . . . Peckham, "Medscape Psychiatrist Compensation Report 2018."

decreasing its flow to those areas . . . Mary O'Hara and Pamela Duncan, "Why 'Big Pharma' Stopped Searching for the Next Prozac," *The Guardian,* January 27, 2016, https://www.theguardian.com/society/

2016/jan/27/prozac-next-psychiatric
-wonder-drug-research-medicine-mental
-illness.

"It is to be hoped that" . . . David Cunningham Owens and Eve C. Johnstone, "The
Development of Antipsychotic Drugs,"
Brain and Neuroscience Advances, December 5, 2018, https://journals.sagepub.com/
doi/full/10.1177/2398212818817498#art
icleCitationDownloadContainer.

psychedelic revival . . . Matt Schiavenz,
"Seeing Opportunity in Psychedelic
Drugs," *The Atlantic,* March 8, 2015,
https://www.theatlantic.com/health/
archive/2015/03/a-psychedelic-revival/
387193.

Even brain stimulation . . . For more on
deep brain stimulation, past and present,
see Frank, *The Pleasure Shock.*

Some techniques involve implanting electrodes . . . Thank you to Columbia psychiatrist Cheryl Corcoran, who shared some
details about her work with deep brain
stimulation in our phone interview on
April 11, 2017.

a variation of the anesthetic ketamine . . .
Benedict Carey, "Fast-Acting Depression
Drug, Newly Approved, Could Help Millions," *New York Times,* March 9, 2015,
https://www.nytimes.com/2019/03/05/

health/depression-treatment-ketamine-fda
.html.

being touted on all the morning shows . . .
"What to Know About Ketamine-Based
Drug for Depression and More," *Today,*
March 6, 2019, https://www.today.com/
video/what-to-know-about-ketamine
-based-drug-for-depression-and-more
-1452994627709.

therapy creates profound changes . . . Eric
Kandel, "A New Intellectual Framework
for Psychiatry," *American Journal of Psychiatry* 155, no. 4 (1998): 457–69, https://
www.ncbi.nlm.nih.gov/pubmed/9545989;
and Louis Cozolino, *The Neuroscience of
Psychotherapy: Healing the Social Brain,*
2nd ed. (New York: W. W. Norton, 2010).

"Psychotherapy is a biological treatment" . . . Eric R. Kandel, "The New Science of the Mind," *New York Times,* September 6, 2013, https://www.nytimes.com/
2013/09/08/opinion/sunday/the-new
-science-of-mind.html.

"One sees as far as one is limited" . . . Niall
Boyce, phone interview, April 19, 2016.

"It's true, [it's like having] a microscope" . . . Matthew State, phone interview, March 13, 2017.

"You're going to see the whole thing" . . .
E. Fuller Torrey, phone interview, January

14, 2016.

"In spite of the fact that state hospitals" . . . Joel Braslow, phone interview, March 10, 2015.

The late neurologist Oliver Sacks agreed . . . Oliver Sacks, "The Lost Virtues of the Asylum," *New York Review of Books,* September 24, 2009, retrieved from https://www.nybooks.com/articles/2009/09/24/the-lost-virtues-of-the-asylum.

Three University of Pennsylvania ethicists . . . Dominic Sisti, Andrea G. Segal, and Ezekiel J. Emanuel, "Improving Long-Term Psychiatric Care: Bring Back the Asylum," *JAMA* 313, no. 3 (2015): 243–44.

"a disgrace" . . . confirmed via emails provided to me by Dominic Sisti on April 29, 2019.

"The debate boils down to one question" . . . Dominic Sisti, phone interview, July 6, 2017.

"The brain is extremely plastic" . . . Maree Webster, interview at the Stanley Medical Research Institute Laboratory of Brain Research, January 14, 2016.

Environmental factors . . . For a great breakdown of the environmental factors associated with developing severe mental illness, see Joel Gold and Ian Gold, *Suspi-*

cious Minds.

antibodies directed against a common feline parasite . . . E. Fuller Torrey and Robert H. Yolken, "Toxoplasma Gondii and Schizophrenia," *Emerging Infectious Diseases* 9, no. 11 (2003): 1375–80, https://wwwnc.cdc.gov/eid/article/9/11/03-0143_article.

schizophrenia found in the Caribbean population . . . Rebecca Pinto and Roger Jones, "Schizophrenia in Black Caribbeans Living in the UK: An Exploration of Underlying Causes of the High Incidence Rate," *British Journal of General Practice* 58, no. 551 (2008): 429–34, https://bjgp.org/content/58/551/429.

Living in cities is linked . . . One of many studies that have shown a correlation between urban life and schizophrenia is James Kirkbride, Paul Fearon, Craig Morgan, Paola Dazzan, Kevin Morgan, Robin M. Murray, and Peter B. Jones, "Neighborhood Variation in the Incidence of Psychotic Disorders in Southeast London," *Social Psychiatry and Psychiatric Epidemiology* 42, no. 6 (2007): 438–45, https://link.springer.com/article/10.1007%Fs00127-007-0193-0.

A two-year government-funded study . . . John M. Kane et al., "Comprehensive

Versus Usual Community Care for First-Episode Psychosis: 2-Year Outcomes from the NIMH RAISE Early Treatment Program," *American Journal of Psychiatry* 173, no. 4 (2016): 362–72, https://www.ncbi.nlm.nih.gov/pubmed/26481174.

"comprehensive, multi-element approach" . . . Thank you to Dr. Robert Heinssen, Dr. Lisa Dixon, and Dr. John Kane for your perspectives on RAISE and early intervention. For more information, see Robert K. Heinssen, Amy B. Goldstein, and Susan T. Azrin, "Evidence-Based Treatment for First Episode Psychosis: Components of Coordinated Specialty Care," National Institute of Mental Health, April 14, 2014, https://www.nimh.nih.gov/health/topics/schizophrenia/raise/evidence-based-treatments-for-first-episode-psychosis-components-of-coordinated-specialty-care.shtml.

people who are troubled by hearing voices . . . For a wonderful examination of voice-hearing, see Charles Fernyhough, *The Voices Within: The History and Science of How We Talk to Ourselves* (New York: Basic Books, 2016), 4.

Yale researchers found that a key difference . . . Albert R. Powers, Megan S. Kelley, and Philip R. Corlett, "Varieties of

Voice-Hearing: Psychics and the Psychosis Continuum," *Schizophrenia Bulletin* 43, no. 1 (2017): 84–98, https://academic.oup.com/schizophreniabulletin/article/43/1/84/2511864.

compared the experience of auditory hallucinations . . . Tanya Marie Luhrmann et al., "Culture and Hallucinations: Overview and Future Directions," *Schizophrenia Bulletin* 40, no. 4 (2014): 213–20.

"Are those cultural judgments" . . . Joseph Frankel, "Psychics Who Hear Voices Could Be onto Something," *The Atlantic,* June 27, 2017, https://www.theatlantic.com/health/archive/2017/06/psychics-hearing-voices/531582.

One popular therapy that takes these cultural judgments . . . For more on open dialogue therapy, see Tom Stockmann, "Open Dialogue: A New Approach to Mental Healthcare," *Psychology Today,* July 12, 2015, https://www.psychologytoday.com/us/blog/hide-and-seek/201507/open-dialogue-new-approach-mental-healthcare.

I saw McLean's version . . . I visited McLean Hospital in August 2017. Thank you to Dr. Dost Ongur and Dr. Joseph Stoklosa for allowing me to visit and for taking time to show me their techniques.

You've heard of the placebo effect . . . For a great discussion of the placebo effect and history, see Jo Marchant, *Cure: A Journey into the Science of Mind Over Body* (New York: Crown, 2016); Melanie Warner, *The Magic Feather Effect: The Science of Alternative Medicine and the Surprising Power of Belief* (New York: Scribner, 2019); and Gary Greenberg, "What If the Placebo Effect Isn't a Trick?" *New York Times,* November 7, 2018, https://www.nytimes.com/2018/11/07/magazine/placebo-effect-medicine.html.

with the psalm Placebo Domine . . . Daniel McQueen, Sarah Cohen, Paul St. John-Smith, and Hagen Rampes, "Rethinking Placebo in Psychiatry: The Range of Placebo Effects," *Advances in Psychiatric Treatment* 19, no. 3 (2013): 171–80.

to attend funerals to "sing placebos" . . . C. E. Kerr, I. Milne, and T. J. Kaptchuk, "William Cullen and a Missing Mind-Body Link in the Early History of Placebos," *Journal of the Royal Society of Medicine* 101, no. 2 (2008): 89–99, https://www.ncbi.nlm.nih.gov/pmc/articles/PMC2254457/.

The word made its way . . . Kerr, Milne, and Kaptchuk, "William Cullen and a

Missing Mind-Body Link."

By the 1960s, the FDA had set . . . Suzanne White, "FDA and Clinical Trials: A Short History," U.S. Food & Drug Administration, https://www.fda.gov/media/110437/download.

saline solution that you believe is morphine . . . J. D. Levine, N. C. Gordon, R. Smith, and H. L. Fields, "Analgesic Responses to Morphine and Placebo in Individuals with Postoperative Pain," *Pain* 10, no. 3 (1981): 379–89.

Parkinson's patients will release dopamine . . . Sarah C. Lidstone, Michael Schulzer, and Katherine Dinelle, "Effects of Expectation on Placebo-Induced Dopamine Release in Parkinson Disease," *Archives of General Psychiatry* 67, no. 8 (2010), https://jamanetwork.com/journals/jamapsychiatry/fullarticle/210854.

"Ultimately it's about being immersed" . . . Dr. Ted Kaptchuk, phone interview, January 18, 2016.

In a study of acid reflux sufferers . . . Michelle Dossett, Lin Mu, Iris R. Bell, Anthony J. Lembo, Ted J. Kaptchuk, and Gloria Y. Yeh, "Patient-Provider Interactions Affect Symptoms in Gastroesophageal Reflux Disease: A Pilot Randomized, Double-Blind, Placebo-Controlled Trial,"

PLoS One 10, no. 9 (2015), https://www.ncbi.nlm.nih.gov/pmc/articles/PMC4589338/.

pushing to rebrand the placebo effect . . . Warner, *The Magic Feather Effect,* 70.

"Each time they tell me" . . . Email to Susannah Cahalan, March 23, 2019.

"If I'd adopted the conventional wisdom" . . . Rossa Forbes, *The Scenic Route: A Way Through Madness* (Rolla, MO: Inspired Creations, 2018), 71. Thank you, Rossa, for sharing your son's story with me over the phone, as well.

Epilogue

"Whenever the ratio of what is known" . . . Rosenhan, "On Being Sane in Insane Places," 397.

"I was surprised initially" . . . Florence Keller, in-person interview, February 18, 2017.

"plays practical jokes on his contemporaries" . . . Julia Suits, *The Extraordinary Catalog of Peculiar Inventions: The Curious World of the DeMoulin Brothers and Their Fraternal Lodge Prank Machines — from Human Centipedes to Revolving Goats to Electric Carpets and Smoking Camels* (New York: Penguin, 2011).

"There is a certain shadowy quality" . . . Lee Ross, in-person interview, February 18, 2017.

"David's fame was based on many accomplishments" . . . A copy of Lee Shulman's speech was provided to me by Lee via email on December 2, 2013.

Jack was thirteen . . . Jack's story of his father and their trip to New York City came from various phone and in-person interviews.

PERMISSIONS

Page 158: Excerpt from Haverford State Hospital medical records. David Rosenhan's private files. Permission granted by Florence Keller and Jack Rosenhan.

Page 170: Excerpt from Haverford State Hospital medical records. David Rosenhan's private files. Permission granted by Florence Keller and Jack Rosenhan.

Pages 193–194: Excerpt from Haverford State Hospital medical records. David Rosenhan's private files. Permission granted by Florence Keller and Jack Rosenhan.

Page 218: Excerpt of questionnaire. David Rosenhan's private files. Permission granted by Florence Keller and Jack Rosenhan.

Page 230: Handwritten excerpt of John Fryer's speech. John Fryer, "Speech for the American Psychiatric Association 125th Annual Meeting," undated, John

Fryer Papers, Collection 3465, 1950–2000, Historical Society of Pennsylvania (Philadelphia). Permission granted by Historical Society of Pennsylvania.

Page 240: Excerpt from yearbook. Stanford University, Stanford Quad, 1973. Print, Stanford University Archives. Reprinted with permission from Stanford University.

Page 287: "William Dickson" medical record. Permission granted by Bill Underwood to publish.

Pages 321–322: Excerpt from Haverford State Hospital medical records. David Rosenhan's private files. Reprinted with permission from Florence Keller and Jack Rosenhan.

Page 325: Excerpt from Haverford State Hospital medical records. David Rosenhan's private files. Reprinted with permission from Florence Keller and Jack Rosenhan.

Page 333: Excerpt from Haverford State Hospital medical records. David Rosenhan's private files. Reprinted with permission from Florence Keller and Jack Rosenhan.

Pages 333–334: Excerpt from Haverford State Hospital medical records. David Rosenhan's private files. Reprinted with permission from Florence Keller and Jack

Rosenhan.

Page 374: Excerpt from Harry Lando, "On Being Sane in Insane Places: A Supplemental Report," Professional Psychology, February 1976: 47–52. Reprinted with permission from Harry Lando.

.

ABOUT THE AUTHOR

Susannah Cahalan is the award-winning, *New York Times* bestselling author of *Brain on Fire: My Month of Madness,* a memoir about her struggle with a rare autoimmune disease of the brain. She lives in Brooklyn.